YORKSHIRE MINERS

BRIAN ELLIOTT

The
History
Press

First published in 2004
This edition first published in 2009

The History Press
The Mill, Brimscombe Port
Stroud, Gloucestershire, GL5 2QG
www.thehistorypress.co.uk

British Library Cataloguing in Publication Data.
A catalogue record for this book is available from the British Library.

ISBN 978 0 7524 4938 8

Printed in Great Britain

Title page photograph: The
Brodsworth Branch banner
in procession during the
Yorkshire Miners' Gala and
Demonstration in Castleford,
20 June 1964. (*NUM/NCB*)

For Arthur Clayton, 1901–2002,
Yorkshire Miner and Local Historian;
and the Yorkshire Area NUM

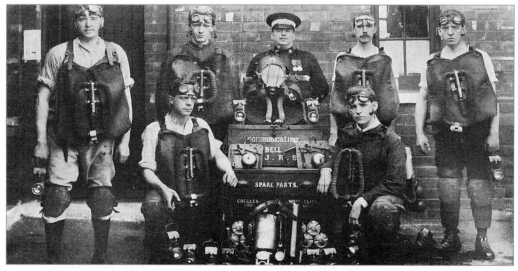

Photographs of rescue teams can be the most compelling of mining images. Here we can
see members of the Wath rescue team in 1923, plus their considerable range of equipment,
including Draeger oxygen apparatus. The men are, left to right, standing: Jarvis, Paskell, Winch
(instructor), Tom Wood and Harry Jones; and kneeling, Whitley and Reg Longworth. The Wath
station (supported by Wath, Manvers, Hickleton, Cadeby and Denaby collieries) was built in
1908, some three years before rescue teams were made compulsory in the Coal Mines Act.
(*NUM*)

CONTENTS

Women cricketers pose for the camera at South Kirkby Colliery sports ground, *c.* 1905. (*Old Barnsley*)

Barnburgh Colliery (later known as Barnburgh Main) was sunk by the Manvers Main colliery company over a three-year period, from 1912. This wonderful photograph probably relates to one of two principal working shafts (nos 5 or 6), the other four being located at the Manvers site. The Barnburgh shafts, which were brick-lined, had to be deep in order to access the Barnsley Seam (at 508yd) and Parkgate Seam (757yd). The bowler-hatted gentleman may have been a director. The large metal bucket, known as a hopper or kibble, was used for removing waste material during sinking operations and also for carrying men up and down the shaft. A large group photograph was also taken, showing at least twenty-five of the itinerant pit sinkers, all wearing their distinctive oilskins and protective hats. (*NUM*)

INTRODUCTION AND ACKNOWLEDGEMENTS

In the 1970s when I was a young teacher in a Barnsley area school, it was an essential requirement to include some aspects of 'Coal Mining' in my environmental studies and geography lessons. For younger children, lessons were brought alive when boys and girls were dressed up in mining gear, and a visit to Grimethorpe pit was an unforgettable experience for us all. This was in the days when we were permitted to tailor our own syllabuses for CSE (Certificate of Secondary Education) to the local area and when the pioneering Geography for the Young School Leaver (GYSL) GCE, which had a large element of course-based assessment, was also available. Nowadays, coal mining is a feature more of history rather than geography in the evolving National Curriculum. Show a group of young children a piece of coal today and there will be some who will not know what it is, where it comes from and what it is used for.

Thirty years ago there were few of us who could envisage that the once great coal industry that powered the Industrial Revolution would be virtually extinct within a generation. A certain young Yorkshire miners' leader did warn of complacency and the ever present threat to jobs and communities. Read the 'Pit Closures' section of Arthur Scargill's pamphlet, *Miners in the Eighties*, and the predictions ring true. At about the same time, the National Coal Board, with its Plan for Coal, was optimistic about the future of the industry. So was Margaret Thatcher. As Conservative leader and Prime Minister she often spoke with optimism about the future of coal and eagerly accepted invitations to visit the new pits. In Yorkshire the great Selby project was fast developing and many old pits were earmarked for an extended life in the context of new technology and massive capital

Commemorative plate showing NCB Barnsley area pits and the 1979/84 'reconstruction' sites. Within a few years every one of these pits and highly-invested new complexes had closed. Only Caphouse colliery remains, as part of the National Coal Mining Museum for the North of England. (*Brian Elliott*)

Hoisting the beam of the large pumping engine at Nunnery Colliery, Sheffield, *c.* 1890. (*Brian Elliott*)

investment. The Prince of Wales Colliery at Pontefract was typical. The new drift and modernisation would mean a life well into the middle of the twenty-first century, according to the NCB. There were also the pit mergers and new complexes focused at locations where mining was a way of life.

There is no doubt that the bitter Miners' Strike of 1984–85 provided the opportunity for an unnecessarily rapid run-down of the British coal mining industry, and the wave of pit closures announced in 1992 virtually sealed this process. The Prince of Wales Colliery closed in 2002 and by the spring of 2004 the Selby Complex (Wistow, Riccall and Stillingfleet) will cease production with a loss of 2,000 jobs. Then there will be only five deep mines left in Yorkshire: Kellingley, Maltby, Rossington, Harworth and Hatfield; and a few others elsewhere. British Energy, responsible for nuclear-powered electricity, has had to have financial aid from the government, so it seems likely that within a few years we will be reliant on gas for electricity generation, most of it imported.

Despite recent events, the history of mining has never been more popular. To some extent this may be due to nostalgia but there is also a general interest in the life and times of miners, their families and communities. Much valuable work is being done to record the memories of former miners, and thank goodness for the National Coal Mining Museum for England at the old Caphouse Colliery site, near Wakefield. Hopefully, this collection of photographs and captions will make a small contribution to our appreciation of the miners of Yorkshire.

Acknowledgements for photograph holders and photographers are given in the form of an italicised credit at the end of each caption.

Several people and organisations gave both help and encouragement for me to complete the book. Philip Thompson at the National Union of Mineworkers offices in Barnsley was kind enough to allow me access to their library and archives. I am also indebted to the former NCB photographer Jeff Poar who trusted me with some of his work. As usual, my friend Chris Sharp of 'Old Barnsley' allowed me access to his extensive Yorkshire postcard collection and I am grateful to Tony Munford and the staff at Rotherham Archives and Local Studies Library for the use of their comprehensive picture library. Sarah Bryce and her colleagues at Sutton Publishing have been very patient and helpful when waiting for me to complete the book. Finally, thanks to my wife and family for putting up with me over the course of the research and writing, especially in recent months.

1

Around the Pit Top

A pit top worker checks the railway lines at Bullcroft Colliery. This view is from a photograph by Scrivens of Doncaster, *c.* 1920s. The mine was sunk at Carcroft between 1908 and 1911, the engineers and sinkers experiencing great difficulty due to the water-laden limestone strata that had to be penetrated. Notice the matching iron-made lattice headframe and well-built colliery buildings. (*Old Barnsley*)

Two images taken from lantern slides dating from *c.* 1890, showing a tub and barrel being lowered down the shaft of a Yorkshire pit during sinking operations. The second photograph shows the timber-lined shaft wall, consisting of wooden planks contained by wedges and iron rails. (*Brian Elliott*)

A rare photograph, showing boring taking place at Wharncliffe Silkstone Colliery (near Tankersley, Barnsley) on 27 August 1914. The machinery, powered by a mobile steam engine, bored to a distance of 127yd to the top of the North Thin Staple. (*NUM*)

Three workmen on the 'National' gas engine at Wharncliffe Silkstone colliery, 22 August 1914. The engine, made by Vickers of Sheffield, was powered by coke oven gas from the pit's own coking plant and served the generators that provided electricity for the colliery. (*NUM*)

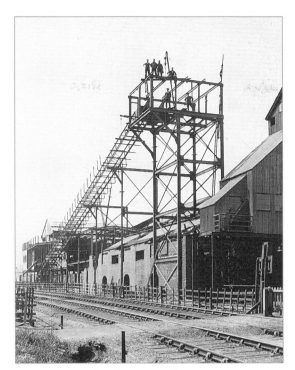

A new iron headgear under construction at Wharncliffe Silkstone *c.* 1910. Note the long metal ramp. The Wharncliffe company chairman, George Blake Walker, introduced many new developments from the 1880s which were way ahead of their time. Innovations included compressed air coal-cutters, electric coal-cutters, miners' hip baths, a coal washing plant, a coke and by-product plant, breathing apparatus and the first ever Mines Rescue station, at Tankersley in 1902. This pioneering pit closed in 1966, but the branch banner rightly claimed its title as the number one branch of the National Union of Mineworkers, Yorkshire Area. (*Brian Elliott*)

John Normansall, who was employed at Wharncliffe Silkstone colliery in the early 1850s, became a notable branch secretary and checkweighman, assisting the miners in disputes over working conditions and wages. He had been a 'pitman' from the age of seven, earning 6*d* a day in Lancashire. When local colliery owners notified their intention of reducing wages by 15 per cent Normansell and others met at the Old White Bear Inn in order to consolidate their interests. The result was the formation of the South Yorkshire Miners' Association. Normansell was one of the leaders in the lock-out of 1864 and was appointed as Secretary of the new South Yorkshire Association, and later Vice-President of the National Union (1870). It was a tragic blow to the South Yorkshire Miners' and the National Union when Normansell died, aged forty-five, in 1875. (*Brian Elliott*)

Opposite above: An early (*c.* 1890) photograph showing pit sinkers (wearing caps) believed to be at Canklow, near Rotherham. Planks and wedges around part of the shaft may be seen in the foreground. Pit sinkers were a 'special breed' travelling from pit to pit to execute their labour and skills, usually living in temporary accommodation over a period ranging from a few months to several years. (*Rotherham Archives and Local Studies*)

Opposite below: Another interesting view of workers who were in the process of sinking a 12ft diameter shaft at Hermit Hill, part of the Wharncliffe Silkstone colliery, 25 May 1915. The shaft reached the Whinmoor Seam at 98yd. (*NUM*)

A curious collection of sailors, miners (and policemen?) at Henry Lodge & Co.'s Ryhill Colliery in August 1919, possibly former employees returning to their place of work after the Great War. If so, their new stint at the pit was a relatively short one since Ryhill Main closed in 1923. In the background we have a good view of the wooden headstock and pulley wheels, reached by a precarious ladder. (*Old Barnsley*)

Opposite: A marvellous early photograph (*c.* 1890) of the wooden headstocks, gearing, ropes and engine house at Rob Royd colliery, a small pit near Dodworth (Barnsley) which became part of the 'Old Silkstone' group (with Dodworth & Redbrook). Note the long ladders, the only way of access to the pulley wheel at the top of the frame. After nationalisation, Rob Royd coal – from the Parkgate and Thorncliffe Seams – was sent underground to Dodworth. (*Michael Hinchliffe*)

Next page: This very interesting group of pit deputies of varying ages appears to have been photographed outside the gates of a grand house after the men had completed a shift at a nearby colliery. It may well have been taken outside Earl Fitzwilliam's Rainborough (also known as Lion) Lodge, so could they be Elsecar miners? And the date? Perhaps just before or after the First World War would probably not be too far out. Of great interest is the variety of lamps, the men's clothing and footwear. Any information regarding the identity of the men and confirmation of the location would be appreciated. (*NUM*)

Another very interesting though somewhat damaged photograph of men and boy miners, from Woolley Colliery, probably dating from 1900 to 1910. Several of the boys look no older than eleven or twelve. Maybe they were mainly surface workers, perhaps from the screens and haulage work. (*NUM*)

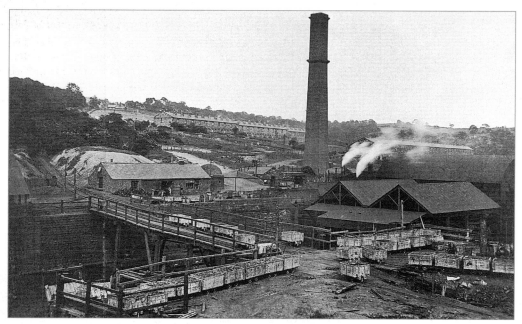

Woolley Colliery: note the miners' houses on the hillside, close to the pit. Records of coal being worked from this area go back to at least the seventeenth century. Woolley Colliery originated from the late 1860s but the modern site had only just been developed – by Fountain & Burnley – when this photograph was taken in *c.* 1914. A £102-million project took place at Woolley between 1979 and 1984. Known as the West Side Complex, a new preparation plant received coal from nearby mines. Despite the massive investment, Woolley closed at the end of 1987. (*Old Barnsley*)

The ceremony of 'cutting the first sod' was recorded in this photograph dating from 23 October 1905. Mrs Thellusson, the wife of the landowner, Charles Thellusson of Brodsworth Hall can be seen at the centre, holding the commemorative spade placed in front of a wheelbarrow. The person arrowed is Henry Goodlad, a miner and checkweighman from the Staveley Iron & Coal Company. He was one of the first employees at the new pit. The Barnsley Seam was reached in October 1907 at a depth of 595yd and production began shortly afterwards. Many of the miners resided in the model village of Woodlands, created by the coal company. (*Alan Paley*)

An interesting photograph by Leonard Scrivens showing the relatively new Brodsworth Colliery, with its no. 1 and no. 2 headgear, in about 1920. The headstocks were of timber so as to withstand better any movement coming from the extracts of coal around the shafts. Brodsworth soon become the biggest pit in Yorkshire and one of the most productive in the UK. In 1957, during 'Bull Week' the men produced 34,422 tons of coal, a record for Yorkshire's 'King Pit'. Mighty Brodsworth closed in 1990. (*Old Barnsley*)

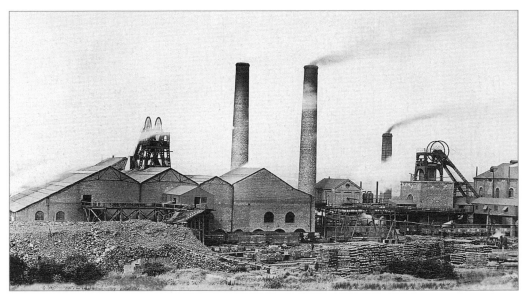

An excellent early view of Grimethorpe Colliery by G.A. Fillingham. Note the lattice headgears. The pit was sunk in a predominantly rural area from 1894 by the Mitchell Main Colliery Company but by completion, in 1897 (delayed due to water problems), the Carlton Main Colliery Company was in control. The Barnsley Bed was reached at 560yds via the no. 1 shaft. The no. 2 shaft, some 80yd distant, also accessed the famous Barnsley Seam. 'Turning the first sod' was a grand occasion, reported in some detail in the *Barnsley Chronicle* of 13 October 1894:

> There was a large and representative company to witness the sod turning on Monday. Those accepting [the invitation] assembled at Cudworth Railway Station early in the afternoon and were conveyed by special train . . . to the scene of operations. The afternoon was fine, the party in good spirits, and there was not a little joking about this 'first trip to Grimethorpe'. A crowd of people from the adjacent villages awaited the arrival of the train. The party having alighted, a move was at once made to the spot where the shafts are to be sunk, indicated by a hoisted flag, and by two large circles marked out on the turf, with a white post at the centre of each. A ring was formed round what will be the down-cast shaft, those present including Mr Joseph Mitchell, Bolton Hall (managing director), and Mrs Mitchell [a long list of VIP's follows]. In the absence of Mr Foljambe [of Osberton Hall, the estate owner], the first sod was cut by Mr G.H. Turner, general manager of the Midland Railway Company. Mr J. E. Mitchell said that they hoped to have an engine-house and all the rest of it erected, and coal to go down to see, within two years from now . . . Before the sod was cut, however, the Revd C.F. Husband had been asked to offer up a prayer.

A third (upcast) shaft was sunk in 1925, eventually reaching the Parkgate and Beamshaw seams. Meltonfield, Haigh Moor, Fenton, Newhill and Thorncliffe seams were also worked in later years. Grimethorpe joined Houghton Main in a complex merger after nationalization and a new power station was in operation by 1960; and six years later the Coalite plant was built. Grimethorpe was part of the £174-million South Side Project (cf. Woolley) during the 1979–84 period, when Barnsley Main, Dearne Valley and Houghton Main linked with Grimethorpe and a state-of-the-art preparation plant was introduced. Despite the huge investment, modernization, output records and being one of the most profitable of the Barnsley area pits, Grimethorpe stopped production in October 1992. (*Old Barnsley*)

Men and boys gather around the pay office at Barrow Colliery, Worsbrough Bridge, photographed by Lamb & Co., early 1900s. The Barrow Haematite Iron & Steel Company was registered in 1864, with interests in iron ore and limestone mining and iron smelting at Barrow-in-Furness. The old Worsbrough Park colliery was purchased by the company and, from the early 1870s, on the strength of the high-grade coking coal that could be exploited, a new colliery was commissioned nearby. (*Old Barnsley*)

Men from the 'last shift' at Barrow Colliery, 17 May 1985. Back row, left to right, they are: Michael Nunn, Andy Lee, Andrew Sykes, Ronnie Charlesworth, Kevin Moxon, Doug Jessop and Tommy Bentley; kneeling, at the front: Danny McBlain, Nigel Myers, Michael Skelly and Terry Barber. From the early 1970s, due to declining reserves, Barrow was increasingly working coal by accessing the old Barnsley Main workings but the latter's reserves were also diminishing by the mid-1980s. Production from Barnsley Main finished in July 1991. (*Brian Elliott*)

An unusual photograph dating from *c.* 1900 and probably relating to Glasshoughton & Castleford collieries, showing a mobile mock-up of an underground haulage machine, complete with pit props, a couple of miners and 'SNAP TIME' in bold letters as a title. One man is in charge of the horse while another takes charge of a small pony and feed bag. The display may have been for a local gala or promotional event. (*NUM*)

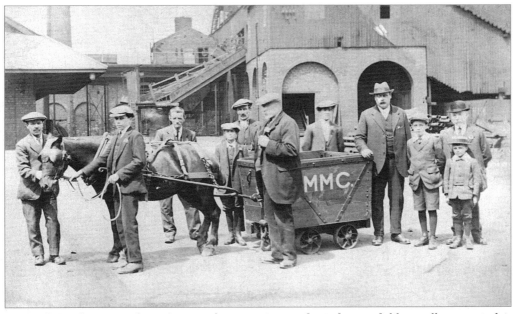

Hospital Sunday was always a popular occasion and, understandably, well supported in Yorkshire's mining communities. Here, a small and smart-looking group at Manvers Main stand by a harnessed pit pony and tub, perhaps preparing for a collection at this important fund-raising event, *c.* 1920. (*Rotherham Archives and Local Studies*)

Programme cover for a 'Grand Open-Air Musical Festival' in aid of Hospital Sunday, Midland Club Field, Royston, 28 August 1921. Yorkshire Miners' President Herbert Smith was the principal speaker, an indication of the great importance of such events. (*Brian Elliott*)

Young children are placed on guided pit ponies at Waleswood Colliery, near Rotherham, *c.* early 1900s. The labels around the horses' necks suggest some kind of competition. Pit pony racing was quite common during strikes. From 1903 pit ponies were not used for underground haulage at Waleswood due to difficulties in recruiting lads as pony drivers. Instead a system of compressed air-powered winches was used for hauling coal from the workings. This photograph may date from the decision to abandon the underground use of the ponies. Notice the small wooden headgear and beaker-type building which may have housed the Capell fan. Despite an underground sit-in and strike, Waleswood was closed by the NCB in 1948. (*Rotherham Archives and Local Studies*)

A group of pit-top workmen (every one wearing a cap) pause for their photograph to be taken, Kilnhurst Colliery, *c.* 1920. Even when overalls were available, most workmen preferred to wear old clothes, including waistcoats. (*Rotherham Archives and Local Studies*)

A special 'Celebration Dinner', housed in a marquee, was held after coal (the Barnsley Seam) was 'found' at Maltby Main, photographed by Scrivens, July 1910. (*Rotherham Archives and Local Studies*)

Pit muck is still on these miners as they wait for a bus to take them home from Monk Bretton Colliery in 1948. Second from the left (at the front) is Jack Lever who worked at the pit for twelve years and lived at Hoyle Mill. The bus stop was opposite the pit, in Burton Road, and the wooden shed in the background was where the miners paid their union subs. The colliery was one of the many victims of the Robens era, closing in 1968. Some of the surface buildings were later used by the NCB as a training centre. (*Coal News*)

Two women from the cast of a local pantomime on a guided visit to Darfield Main in 1966. The guides are John Reeves (left) and Neil Robinson. (*Malcolm Robinson*)

Two veteran miners, one wearing clogs, at the pit top after finishing their shift at a Yorkshire colliery, *c.* 1930s. (*NUM*)

Six miners assemble at the pit top after completing the last shift at Wharncliffe Woodmoor 4 and 5 colliery (formerly known as Carlton Main) at Carlton, near Barnsley in July 1970. The group includes Brian Shore (extreme left, upper photograph and centre background in the lower image), deputy Bob Ashton (wearing spectacles on both photographs) and Bernard 'Boz' Rowley (fifth from left, upper and second left, lower). Bernard started work as a lad at Brierley colliery before moving to Grimethorpe and then Wharncliffe Woodmoor 4 and 5. Wharncliffe's sister pit, Woodmoor 1, 2 and 3 had closed four years earlier. (*Thelma Carnevale*)

A historic moment at Rockingham Colliery: the last tub of coal came from the Low Fenton Seam, wound up the no. 2 shaft, on Tuesday 20 November 1979. (*Sheffield Morning Telegraph*)

Miners from Rockingham celebrate the closure of their 104 year-old pit at Christmas 1979. Veterans in the photograph are, left to right: Bill Beardsall, Walter Boreham, Bill Keeling, Norman Happs, -?-, Jackie Hobson, Bill Shackley, Johnny Mallinson, Arthur Clayton and Arthur Bennett. Rockingham was noted for having long-serving employees, many working for half a century or more. (*NUM*)

Coke oven workers at Monckton in 1908. The Monckton Coke and Chemical Company continues to exist as part of UK Coal's operations but the Monckton colliery complex closed in 1966. My paternal grandfather worked at the pit and my maternal grandfather was employed at the chemical works. These men are assembled in front of a corrugated building known as The Ram which housed machinery. Many of them and their families would have been new migrants into the area. The only individual identified is the bearded man standing on the extreme left who is believed to be James Wood. He lived in a terraced house at Strawberry Gardens in Royston, a neighbour of my great-grandfather, James Winter who was a shoemaker. (*Brian Elliott*)

After the 1984–85 miners' strike, production records were broken at numerous Yorkshire pits. Here is a celebratory example from Darfield Main in 1986. Darfield was merged with Houghton Main and production ceased in the summer of 1989. (*Old Barnsley*)

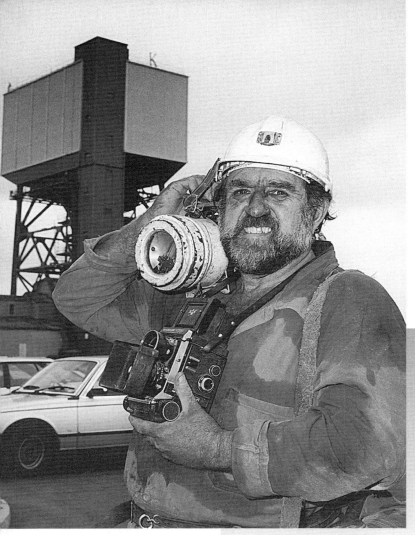

Jeff Poar worked as an NCB photographer, based at Coal House in Doncaster. From a South Wales mining family, he recorded thousands of pit top and underground scenes from the 1970s until the late 1980s. This photograph, on the pit top at Kellingley, dates from the late 1980s. (*Jeff Poar*)

Gary Goodlad sitting astride the concrete plug that marks the site of one of the shafts of Brodsworth Colliery, 1991. Gary worked as a young apprentice at Brodsworth in the early 1960s, became a deputy engineer at nearby Markham Main at Armthorpe in 1970 and in recent years has been employed by HM Inspectorate of Mines. Gary is the grandson of Brodsworth miner Henry Goodlad (see page 17) who stood on this site during the sod-cutting ceremony in 1905. (*Alan Paley*)

2

Underground

The early use of steel roof supports is evident in this lantern slide, dating from about 1890, used as an educational aid on mining courses in the West Riding during the 1920s and 1930s. Produced by Firth's of Sheffield. (*Brian Elliott*)

Three illustrations taken from *The Safe Use of Explosives in Coal Mines*, published by HMSO in 1931.

The first example (above left) concerns 'Charging the Shot-Hole'. We can see the shot-firer pushing the primer cartridge into the hole with a tool called a stemmer. The quantity of explosive used was said to be a matter of judgement of the shot-firer, based on practical experience. This second illustration (above right) shows the shot-firer making a final check for firedamp, a vital task carried out within a radius of 20yds of the shot-hole. The shot-firer holds the lamp in the suspected atmosphere, and lowers the flame until a feint line of blue is seen above the yellow centre. A luminous halo would be seen above the small lamp flame if firedamp was present. The percentage present was judged according to the size of the 'gas cap'.

The final photograph shows firing from shelter. The shot-firer makes sure that all persons working in the vicinity have taken shelter, as well as himself. Here we can see him taking the handle of the exploder from his pocket.

A superb atmospheric self-portrait by Irvin Harris (1912–98), a collier and keen amateur photographer, who took a series of 'unofficial' underground scenes at Woolley Main and Barrow Colliery. This early example dates from *c.* 1937 and entailed the use of a long exposure in the absence of flashlight, the only light source coming from his cap lamp. (*Harris Collection*)

Another excellent underground photograph from the camera of Irvin Harris. It shows Arthur Hudson (1904–79), a Barrow Colliery deputy, making notes, *c.* 1950. (*Harris Collection*)

A large group of deputies assemble at the pit bottom, Maltby Main, *c.* 1930. In places, the roof does not appear to be very sound. (*Rotherham Archives and Local Studies Library*)

Another interesting photograph showing deputies at Maltby Main, *c.* 1930. Here, we get a good view of the timber roof supports, measuring a good 6ft in height, short wedges of timber connecting the top of the props with the roadway roof. (*Rotherham Archives and Local Studies Library*)

Undated photograph showing tubs 'parked' at the well-lit pit bottom, near the cage, at Wharncliffe Silkstone Colliery (Pilley, near Barnsley), one of the oldest of the great Yorkshire pits. The coal produced excellent coke and was rich in by-products. By the early 1950s the surviving shafts were used only for ventilation and pumping, coal from the Fenton Seam being transported to the surface via a drift. (*Keith Hopkinson*)

The well-kept and whitewashed stables at Wharncliffe Silkstone, the proud horse-keeper in the distance. The pony nameplates for 'Duke', 'Victor' and so on are just visible. Note the neat stone setts and rails on the ground. (*A.K. Clayton*)

The pit pony stables at Woolley Colliery, *c.* 1950. The cat was a great asset in order to keep down vermin. Since the 1911 Coal Mines Act, horse-keepers had to keep a record of their ponies (each one named), and monitor their condition and daily movement (see chalk board on upper right of the photograph). There are many amusing stories about pit ponies, including the case at Darfield Main when hungry Bruce ate a miner's false teeth which had been wrapped in paper, placed in a jacket pocket and hung up. This animal had a well-deserved reputation for gobbling snap at any opportunity. (*Harris Collection*)

The famous (but notoriously gassy) Barnsley Seam at Brodsworth Colliery's No. 8 Unit, looking towards the Main Gate, 13 February 1956. Mechanised loading at the coalface had been introduced three years earlier. This photograph was taken at a time when Brodsworth was breaking production records, and a major modernisation scheme had started. The colliery was able to promote itself as 'the biggest pit in the world' and build on its reputation as Yorkshire's 'King' pit. By 1957 the pithead baths could cater for 4,000 men, a measure of the size of the colliery. (*NUM*)

Arthur Clayton controlling the Rockley Tandem Bunker at Rockingham Colliery, *c*. 1964. When he retired, in 1966, Arthur had worked for over fifty years as a miner at Wharncliffe Silkstone and Rockingham collieries.
A much respected and very popular local historian, Arthur was awarded the BEM for his contribution to English heritage, a public acknowledgement of his many years of research, writing and teaching. He died, aged 101, in 2002. (*Hugh Wood & Co.*)

The Rockley Tandem Bunker at Rockingham was manufactured and installed by Hugh Wood & Co. of Gateshead (*Hugh Wood & Co.*)

Lord Robens of Woldingham (left) and face-worker Des Spencer in Riddings Drift, Shafton Seam, 8 August 1970. Alfred Robens was appointed as chairman of the NCB nine years earlier, in 1961, when some political commentators saw him as a future party leader. He continued in office until 1971, overseeing a massive rundown of the industry, when several hundred pits closed, but there was no national strike. Riddings was a new drift mine which had opened in 1969 and was accessed via a surface entrance in South Kirkby pit yard but linked to Ferrymoor Colliery at Grimethorpe. Riddings, or Ferrymoor/Riddings as it became known, soon set coal production records, using the latest technology and 'American-style' short-faces. Des Spencer had moved to the new pit following the closure of Wombwell Main in 1969. (*Nick and Janet Kenworthy*)

A ladies group are shown in this photograph, along with their escorts and guides, on an underground visit to Manvers Main in 1974 or 1975. Standing, left to right, are Fred Parks, Jean Tindle, Betty Corker, Jean Fletcher, Albert Blessed, Frank ?, -?-, -?-, Freddie Smith (training officer) and -?-; the seated ladies include Jackie Marshall (third from left). In the postwar years Manvers and its associated coal preparation and coking plants was the centre of a huge modernisation scheme involving links between Barnburgh, Kilnhurst and Wath in what became known as the Manvers Complex. Closure came in March 1988, followed, in the 1990s, by a massive regeneration of the derelict site and new road infrastructure. (*Jean Fletcher*)

Inauguration by
H.R.H. Prince of Wales
of The Prince Charles Drift
Prince of Wales Colliery
Pontefract
Wednesday 25th June 1980

A young Prince Charles, on a special visit to open the new Prince
Charles Drift at the Prince of Wales Colliery, Pontefract in 1980. Note
the white overalls! (*Jeff Poar/NCB*)

The official programme for the inauguration of The Prince Charles
Drift at the Prince of Wales Colliery, 25 June 1980. The main seams
in the new drift were Castleford Four Foot and Warren House; also
worked were the Haigh Moor and Flockton Thin Seams, providing
reserves 'well into the next century'. The pit closed in 2002. (*NCB*)

Another royal visitor to a Yorkshire pit, and always a very popular one, was the Duchess of Kent, wearing a headscarf under her helmet. The entire VIP party are wearing white overalls for the special occasion. On 29 October 1976 it was the Duchess of Kent who started the drill rig on a field outside the village of Wistow to inaugurate work on the first mine shaft of the new Selby Coalfield. (*Jeff Poar*)

A superb photograph by Jeff Poar showing hand-filling on the 16in-high Beeston Seam at Emley Moor Colliery in January 1983. Veteran miner Reuben Kenworthy is seen in the old process of 'hand-filling', using a hammer-pick (known as a 'tommy 'awk' at Emley and a 'peggy' at some pits) to loosen coal that had escaped the controlled shot-firer's blast; he would then shovel the coal on to a moving conveyor. The power would come from his arms, chest, and stomach and care must have had to be taken not to dislodge the props with his legs due to the cramped conditions. Over 2,500 shovelfuls per shift! The 'Sixteen Tons' song of Tennessee Ernie Ford comes to mind. At the age of fifty-nine, Reuben averaged an amazing 22 tons per shift by this manual process, much preferring it to work on a mechanized face. Note the rough-hewn timber roof support and split-bar placed there for safety by Reuben's 19-year-old son, Chris. The stubby props, specially imported, give off an early-warning creak in advance of a roof-fall. A legendary Polish collier, Marion Rokicki, or 'Rocky', worked the same face and moved almost forty-tons on one shift, the pit record. At the time, Emley Moor, located between Huddersfield and Wakefield, was the NCB's smallest mine but produced very high-quality coal. (*Jeff Poar*)

Jimmy Dane bores into the low face at Emley Moor with an air-drill, prior to setting explosives to loosen the coal. The little Emley pit closed in 1985. (*Jeff Poar*)

Brian Saunders, driving a powerful shearer at Rockingham Colliery, a contrasting scene to the previous two photographs. Notice the huge teeth on the machine which could generate a great deal of dust. (*Jeff Poar/NCB*)

Modern, powered hydraulic supports being adjusted in a Yorkshire pit, early 1980s. They are capable of 'self-advancement' and safer than the first generation of hydraulic supports. Below, the men show two large lumps of coal. (*Jeff Poar*)

A Shearer/power loader being sprayed to keep dust down on this composed photograph, taken in the Beamshaw Seam at Glasshoughton Colliery, 1980s. The men are Brian Henderson (front) and Les Evans. The cowl covering the Shearer saves a great deal of cleaning of the huge machine. The conveyor has stopped working as there should not be such a build-up of coal. (*Jeff Poar*)

A lovely photograph by Jeff Poar of a group of 'Shearer/power loader men' in a Yorkshire colliery, early 1980s. Again, we can appreciate the massive quality of the machine and the cramped conditions of work. (*Jeff Poar*)

A new roadway/heading under development by mechanised boring in a Yorkshire pit. We also have a good view of the arched girders and the 'air-bag', a conduit for dust extracted via the retractor fan. (*Jeff Poar/NCB*)

Electrician Roy Spilsbury (left) and fitter Ray Miller hard at work at a Yorkshire colliery face. (*Jeff Poar/ NCB*)

Good rapport between photographer and subjects is apparent in this underground scene, dating from about 1980. It would be pleasing to hear from anyone who can recognise any of the miners in this group. (*Jeff Poar*)

Another example of the use of modern hydraulic (Dobson) props in a Yorkshire pit. Note the iron mesh between the roof spaces. (*Jeff Poar/NCB*)

A photograph by Jeff Poar of a 'pit paddy' taking miners to their places of work. Again, it would be interesting if anyone could identify any of the people and also the location. (*Jeff Poar/NCB*)

An underground scene from the new Ricall mine, part of the Selby Complex, late 1980s. A deep seam is being made ready for production. (*Jeff Poar/ NCB*)

A power loader at work on a thick seam in the Selby complex, at Whitemoor Colliery. (*Jeff Poar/ NCB*)

A huge drivage heading machine pauses from its work on a wide underground roadway at Thorne, the most easterly colliery in Yorkshire, 1987. The boom type of machine has a rotating head fitted with picks and would be capable of cutting out a rectangular profile. This pit had recently merged with Hatfield via a management consortium but in 1988 it was 'mothballed'. (*Jeff Poar/NCB*)

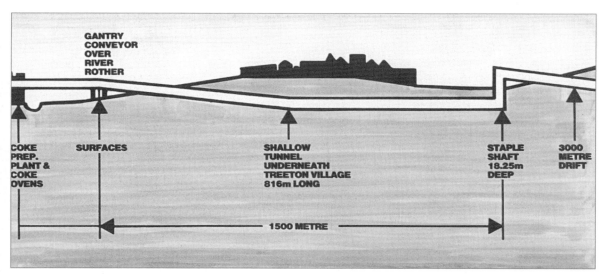

This graphic illustrates the coal transportation system in operation at Treeton Colliery, near Rotherham, developed between 1977 and 81. It enabled coal to be transported 'without seeing the light of day' from the new drift on a conveyor and then discharged into a staple shaft. From there another conveyor, housed in a long shallow tunnel passing under the village, took the coal to a preparation plant at Orgreave Colliery, half a mile away; and from there to the coke ovens of the British Steel Plant. Despite the massive investment, Treeton closed at the end of 1990. (*NUM*)

3

Years of Struggle

A group of miners take a rest during outcropping (digging for coal in shallow workings) at Midhope, near Sheffield, probably during the 1912 strike. They were still capable of a great deal of humour, chalking 'Shag's Main' on the side of an impromptu wooden box suspended by a rope and pulley over the 'pit' – and also inscribing the same on an old tin bath. What appears to be a sack of coal suggests that it was a worthwhile activity. (*Old Barnsley*)

A previously unpublished architect's sketch of the new South Yorkshire Miners' Offices in Barnsley, produced by local 'photographic artists' Wilson & Bullock. This distinctive building is now regarded as the first purpose-built trade union headquarters in the world. At the national conference of mineworkers held here in 1874, a few days after the official opening, John Normansall spoke with great pride about the 'cheerful and pleasant' new premises but also referred to the dreadful working conditions, injuries and loss of life that was wrecking so many mining families. (*NUM*)

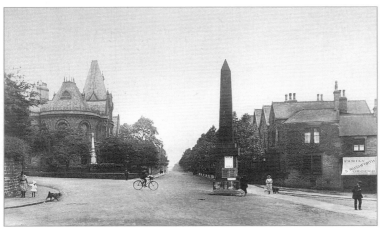

This Edwardian postcard view of the miners' offices, helps us to appreciate its prime location, at the corner of the fashionable Huddersfield and Victoria Roads and overlooking the Mence Obelisk which marked the entry into the town from Old Mill Lane via Church Street. (*Brian Elliott*)

A modern, 1996, view of the miners' offices which in recent years has served as the headquarters of the National Union of Mineworkers. Architecturally, it is one of the town's grandest Victorian buildings and has been the setting of so many important meetings and gatherings over a 130-year history. Despite pit closures, its small staff and volunteers continue to serve the needs of thousands of miners, former miners and their families. (*Brian Elliott*)

JANUARY, 1896

STANDARD LIST OF PRICES

TO BE PAID FOR VARIOUS KINDS OF WORK AT THE

WHARNCLIFFE SILKSTONE COLLIERIES

TANKERSLEY, NEAR BARNSLEY.

AS FIXED BY AGREEMENT AND ARBITRATION.

GENERAL.

	Standard Rate. s.	d.	
Horse Drivers … … … from 1s. to	2	6	per day.
Ordinary Labourers … from 3s. 8d. to	4	2	,,
Colliers Labouring … … … …	4	2	,,
Colliers Lab'ring providing their own Tools	4	6	,,
Woodtaking (piece work) … … …	0	1½	per ton.
Bars (setting bars with props) … …	0	4	per bar.
,, (setting bars and cutting holes) …	0	6	,,
Waterlading (under 100 yards) … …	0	2	per tub.
,, (over 100 yards) … … …	0	3	,,
Filling Dirt … … … … … … …	0	2½	,,
Emptying Dirt … … … … … …	0	2½	,,
Hurrying Rises, bord, over 40 yards …	0	0½	per ton.
,, ,, slant, ,, 60 ,, …	0	0½	,,
,, ,, end ,, 80 ,, …	0	0½	,,
Wheel Holes, 3 feet, 6 inches deep… …	2	6	each.
,, ,, 5 ,, deep … … …	3	3	,,

SILKSTONE SEAM.

	s.	d.	
COALS, in headings, unriddled … …	1	4½	per ton.
,, in benks, on the bord, unriddled	1	4½	,,
,, ,, ,, ,, riddled …	1	6½	,,
,, ,, on the end, unriddled	1	5½	,,
,, ,, ,, ,, riddled …	1	7½	,,
* ,, ,, ,, ,, holed, unrd'ld	0	11½	,,
* Fixed by Judge Bedwell, Q.C., Arbitrator. *Price at present in force* … … … …	1	1	,,
,, ,, ,, ,, ,, riddled	1	1½	,,
HEADING, bord, 5 feet wide … …	3	7	per yard.
,, ,, 7 ,, ,, … …	3	10	,,
,, end, 5 ,, ,, … …	3	10	,,
,, ,, 7 ,, ,, … …	4	0	,,
,, slant, 5 ,, ,, … …	3	10	,,
,, ,, 7 ,, ,, … …	4	0	,,
CUTTING ENDS in leading benks (2 fast ends) from 30 yards to 20 yads wide	0	9	per yard.
,, ,, 20 ,, ,, 12 ,, ,,	1	0	,,
,, ,, 12 ,, ,, 6 ,, ,,	1	3	,,
,, ,, under 6 yards wide …	1	6	,,
GETTING UP seat coal in faces (where required)	0	2	,,
PACKING, 4 feet, 6 inches wide … … …	1	2¼	,,
,, when over 5 feet, 9 inches high	1	3¼	,,
,, 6 feet wide … … … …	1	7	,,
CUTTING Switch … … … … …	0	6	,,
RIPPING (Silkstone), 6 feet x 2 feet …	6	0	,,
,, ,, 8 ,, x 2½ ,, … …	8	0	,,
,, ,, 9 ,, x 3 ,, … …	9	6	,,

WHINMOOR SEAM.

Fixed by T. Marshall, Esq., Umpire, 1891.

	Standard Rate. s.	d.	
COALS, in headings, unriddled … …	1	10½	per ton.
,, in benks, ,, … …	1	10½	,,
,, ,, ,, holed …	1	6½	,,
,, ,, ,, holed …	1	8½	,,

Where levels are opened out by wide places instead of headings, heading prices to be paid.

	s.	d.	
,, extra for tramming on face over 30 yards	0	0½	per ton.
HEADING, bord, 5 feet wide … …	4	9	per yard.
,, ,, 7 ,, ,, … …	5	0	,,
,, end 5 ,, ,, … …	5	0	,,
,, ,, 7 ,, ,, … …	5	3	,,
,, for extra width, 3d. per ft. extra.			
TAKING SIDE OFF (when ordered):—			
bord, 1 inch to 18 inches … …	0	9	,,
,, 18 ,, ,, 36 ,, … …	1	0	,,
end, 1 ,, ,, 18 ,, … …	1	0	,,
,, 18 ,, ,, 36 ,, … …	1	3	,,
BUILDING MIDDLE PACKS, 4 ft. 6 in. wide	1	0	,,
extra for every foot in width … …	0	3	,,
DINTING, 5 ft. wide per vertical. inch	0	1½	,,
,, 7 ,, ,, ,,	0	2	,,

PARKGATE SEAM.

	s.	d.	
COALS, in headings, unriddled … …	1	5½	per ton.
,, in benks. ,, … …	1	5½	,,
,, ,, riddled … …	1	7½	,,
HEADING, bord, 5 feet wide … …	4	5	per yard.
,, ,, 7 ,, ,, … …	5	0	,,
,, end, 5 ,, ,, … …	5	0	,,
,, ,, 7 ,, ,, … …	5	8	,,
,, slant, 5 ,, ,, … …	5	0	,,
,, ,, 7 ,, ,, … …	5	8	,,
,, sideloose … … … …	2	6	,,
RIPPING GATES, 7 feet wide … … …	2	6	,,
PACKING, 4 feet, 6 inches wide … …	1	2¼	,,
,, 6 feet wide … … … …	1	7	,,
CUTTING Brazels … … … … …	0	6	,,

THIN SEAM.

Fixed by Arbitration, 1896

	s.	d.	
COALS, in headings, unriddled … …	2	2	per ton.
,, in benks, unriddled, 2 ft. 6 in. thick	2	2	,,
Rockley District. ,, ,, ,, 2 ft. 6 in. to 2 ft. 7 in. ,,	1	11	,,
Rockley District. ,, ,, ,, 2 7 ,, 2 9 ,,	1	9	,,
Rockley District. ,, ,, ,, 2 9 ,, 3 0 ,,	1	8	,,
Rockley District. ,, ,, ,, 3 0 and upwards …	1	7	,,
,, ,, riddled, 3d. more than the above prices.			
,, ,, holed, 5d. less ,, ,, ,,			

HEADING, bord (coal only)	3/0 per yard	5	feet wide.
,, ,, ,,	3/9 ,,	8	,,
,, end ,,	3/3 ,,	5	,,
,, ,, ,,	4/0 ,,	8	,,

Where levels are opened out by wide places instead of headings, heading prices to be paid.

DINTING, 1d. per inch thick per yard, to be paid for taking up bottom (dirt or seat coal) in headings or gates 5 feet wide, and 1½d. per inch 8 feet wide.

PACKING, 1/- per yd., 6 ft. wide; 1/3 per yd., 8 ft. wide.

Rockley District. DINTING, 1¼d. per inch thick per yard, to be paid for taking up bottom in headings or gates 5 ft. wide; and 1¾d. per inch 7 ft. wide, including 2 yards of pack 6 ft. wide.

Rockley District. PACKING, 1/0 per yard, 4 ft., 6 inches wide; 3d. per yard for each additional foot in width.

All Wages must be Billed at the above rates, and all excess rates entered separately.

GEO. BLAKE WALKER, MANAGING DIRECTOR.

NORTHEND, PRINTER, NORFOLK ROW, SHEFFIELD.

Typical 'Price List' issued by Wharncliffe Silkstone Collieries in 1896 'as fixed by agreement and arbitration'. These remarkably detailed rates of pay lists were issued on a regular basis by the private colliery companies. The union had to fight long and hard over unfair and divisive wage rates and piecework pay, including the notorious 'butty' method. My grandfather used to have to meet his 'paymaster' in a pub yard when the week's money was shared out. (*NUM*)

The Elliott family of Deepcar, my direct ancestors, assemble for a studio photograph in *c.* 1891–2. My great grandfather, Jonas (seated), was dying from the respiratory disease brought about by working in local coal and ganister pits. He was just forty-seven years old when he died. His eldest surviving son, George (centre, standing) was only thirty when he succumbed to the same 'complaint'. Ganister was the raw material mined from the local hard sandstone in places such as Deepcar and Loxley, near Sheffield – in great demand for lining furnaces. The hidden human cost was tremendous and has been almost forgotten. The men and boys who extracted the stone did so in dreadful conditions, most of them suffering from chest-related diseases, dying young and leaving their families in poverty. The boy on Jonas's knee is his son, Fred Elliott (my grandfather, see page 122) who was also to have a lifetime in mining, dying at the age of sixty. (*Brian Elliott*)

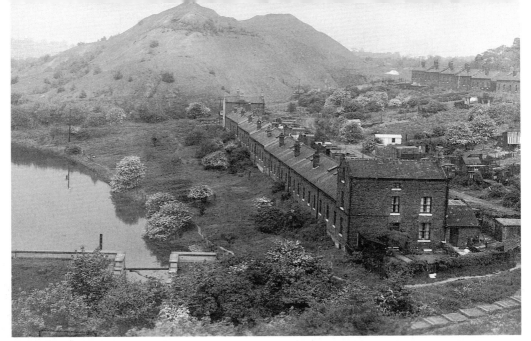

Westwood Row or Rows was a small settlement at the edge of Tankersley parish, built to house non-union labour during the 1869–70 mining dispute which dragged on for seventeen months. The Newton Chambers company, who owned five pits, had reduced employees' wages and refused to negotiate with the miners' union. In the context of an earlier dispute, a lockout and recruitment of non-union labour, it is not surprising that there was great bitterness and unrest from the dismissed miners. The climax came on the morning of 21 January 1870 when several hundred angry men assembled at the Westwood Rows cottages, some attacking the properties. The event attracted national newspaper coverage. There were no fatalities but twenty-three men were arrested and eleven were subsequently gaoled. The background and events of the 'Thorncliffe Riots' is told in Mel Jones's article in *Aspects of Barnsley* (Wharncliffe Books, 1993, reprinted 2003), edited by Brian Elliott. Westwood Rows was demolished in 1965, shortly after this photograph was taken. (*Brian Elliott*)

Miners extracting coal from Clarkson's brickyard, Warren Quarry Lane, Barnsley during the bitter 6-week 'minimum wage' strike of 1912. Almost all British pits closed. The hurriedly assembled Coal Mines (Minimum Wage) Act of 29 March 1912 did not satisfy all the miners' demands, though face-workers achieved a minimum rate of pay, albeit through district rather than national agreement. Through strain and overwork during the complex negotiations, the miners' President, Enoch Edwards, died, on 28 June 1912. (*Brian Elliott*)

Police and armed forces from distant areas were drafted into Yorkshire pits and communities during the 1893 lockout when miners refused to accept a 25 per cent wage cut. This certainly happened in the autumn of 1893 at several Barnsley area locations, including Woolley Colliery, as shown in this carefully composed photograph. (*NUM*)

A penny leaflet relating to the 'Featherstone Massacre' on the evening of 7 September 1893 when a contingent of South Staffordshire Infantry was ordered to open fire on a demonstrating crowd. Two men were injured in the first round and two miners, James Gibbs and James Duggan, were killed in the second volley. (*NUM*)

BULLETS for BREAD!

✳ ✳ ✳

THE

FEATHERSTONE

MASSACRE.

WE WOULD RATHER
BE SHOT DOWN
THAN HUNGERED
TO DEATH.

One Penny.

Miners tied to properties owned by coal companies could face the prospect of eviction during strikes and disputes. This was the case during the so-called Bag Muck Strike at Denaby Main on 6–7 January 1903. The late Jim McFarlane's book, *The Bag Muck Strike 1902–1903* (Doncaster Libraries, 1987) provides us with a well-researched account of the dispute and all the evictions. (*NUM*)

Doncaster Road, Denaby Main, in about 1910. (*Old Barnsley*)

The evictions that took place at Kinsley in 1905 were photographed by Wales of Hemsworth who produced a fine series of postcards. Here are three examples showing the evicted miners and their possessions, children sleeping in temporary accommodation thanks to Thomas Elstone, landlord of the Kinsley Hotel and a property in the process of eviction. More details can be found in Pat and Renee Pickles's book *Kinsley Evictions 1905*. (*Old Barnsley*)

The funeral of 'some of the victims' of the Barrow Colliery disaster, Worsbrough village, 19 November 1907. Seven men were killed and nine others were injured in a horrific cage accident. Tom Cope, Byas Rooke, Isaac Farer, Walter Goodchild, T.W. Jennings, Frank Dobson and C. Adams were the unluckiest of the ascending miners, thrown out of a double-deck, open-fronted cage, falling to the depth of a 200ft shaft on 15 November 1907. (*Brian Elliott*)

The eleven miners killed in an explosion at Wharncliffe Silkstone Colliery, near Barnsley, on 30 May 1914 were commemorated in a postcard produced by Warner Gothard. The pit was said to be regarded 'by Mining Experts to be one of the safest in South Yorkshire'. Many more deaths would have occurred but for the shift being shorter than normal because it was Whit Saturday. (*Brian Elliott*)

'An anxious crowd' gathers at Thrybergh Hall pit (Kilnhurst, near Rotherham) for yet another colliery lockout during the early years of the twentieth century, 5 January 1906. (*Rotherham Archives and Local Studies Library*).

A large group of men, boys and a single policeman, on an outcrop site near to a small Barnsley pit called the Penny Duck Colliery during the 1912 strike. (*Old Barnsley*)

Another compelling photograph showing men (and a small girl) outcropping coal, using picks and shovels, in the Dodworth area of Barnsley, probably during the 1912 dispute. Notice the neck scarves that all the men are wearing. The man next to the girl is Mr Hinchliffe and the girl is his daughter, Neruda. (*Michael Hinchliffe*)

This Salvation Army soup kitchen in Barnsley obviously did stirling work during the 1921 miners' strike, serving up to 500 free dinners every day. The salvationists, consisting of five men, a boy and four women, display jugs, a ladle, a mug and an essential loaf of bread for the occasion. (*Brian Elliott*)

Plates and baskets of sandwiches: Jump Distress Committee assemble for a local photographer, possibly J.R. Short, during the 1926 miners' strike. Back row, left to right: Fred Pantry, Herbert Broadbent, Mr Whorton, Mr Pettinger, Billy Hyde, John Lee, Wilf Boyd, Frank Evans, Louis Green, Mr Hatton, Walter Evans and Noah Hill; middle row: ? Whorton, Bertha Pantry (née Boyd), Edna Lee (née Boyd), ? Swallow, Lily Hyde (née Parr), -?-, Florrie Sanderson (née Allen), Mrs Hill, Mrs Jackson, Betty Evans, Mrs Sharp, Mrs Orwin, Annie Green (née Jolly); front row: -?-, Gertie Clifton, Edith Evans, Mrs Beckett, Harry Haywood, Councillor Reg Preston, Maud Preston (nee Boyd), Mr Bott, Mr Orwin, Mrs Melling, -?-, Mrs Sidebottom and Mrs Carsley. (*Brian Elliott*)

A group of widows, relatives and union officials (including President Herbert Smith, standing, fifth from left) at Maltby, following the disaster of 28 July 1923 when twenty-eight men lost their lives. (*NUM*)

Herbert Smith (1862–1938), the popular, and usually flat-capped, miners' leader. Born in the workhouse and a pit boy at the age of ten, he was President of the Yorkshire Miners' Association from 1906 and a familiar figure in Barnsley and Yorkshire. His honest, down-to-earth and no-nonsense views were given to a host of politicians, including Baldwin. He died at his desk in the Barnsley offices. (*Brian Elliott*)

Issuing bread from the Providence Club at Darfield, near Barnsley, during the 1926 strike. I have spoken to elderly people who remember the places available for food and sustenance during the strike but, despite the extreme hardship, some were too proud to attend. (*R.J. Short*)

Picking coal at Tankersley Park during the 1926 dispute. The man in the trench is 'Pooler' Lee and Dennis Bedford is in the foreground. Others known to be in the picture but not now precisely identified include Harry Chalk, Boscoe Hardy and Mr Woodhead. (*Brian Elliott*)

Another familiar scene in many parts of South Yorkshire during the 1926 strike: men and women with children outcropping coal, in the waste by Carr House Colliery in the Rotherham area. (*Rotherham Archives and Local Studies Library*)

Lest We ✝ Forget

In Memoriam

In affectionate memory of the following, all of whom lost their
lives in the Wharncliffe Woodmoor Colliery Disaster,
on Thursday, August 6th, 1936.

John William Harold Abbott   Benjamin Hodgson
Walter Allott   John Edward Hopes
Cleasby Bailey   Enoch Hulson
William Arthur Bateman   Charles Edward Ismay
Arthur Bird   John Jackson
Henry Birkhead   John David Jones
Lewis Boyd   Samuel Kirk
Alfred Brown   Henry Lee
John Brown   James Robert Miller
Samuel Brown   Owen Owens
William Buckley   Hiram Clarence Parkin
John Bullington   James William Poole
Cecil Chapman   William Proctor
Victor Clarkson   John Roscoe
Frederick Cooper   Harold Rowe
Ernest Dalby   Ernest Scargill
John Donelly   William Henry Senior
Walter Duerden   Joseph Thomas Smith
William Alfred Ellis   Walter Smith
George Farmery   Alexander George Henry Thompson
John Fletcher   George Thompson
Irvin Foster   William Alfred Tompkins
James Green   Herbert Travis
Richard Brookes Grimshaw   John Waugh
Frank Hadfield   Archie White
Arthur Molineaux Haigh   William Whiteley
Herbert Hall   George Henry Wilson
Harry Hatfield   Harry Wright
Horace Llewellyn Hepworth   Richard Wright

County Borough of Barnsley

Wharncliffe Woodmoor Colliery Disaster

✝

In Memoriam

United Memorial Service

TOWN HALL, BARNSLEY
at 3 p.m. on

Thursday, August 13th, 1936

conducted by

THE LORD BISHOP OF DERBY

and

REV. F. LUKE WISEMAN, B.A.

(Ex-President, Methodist Church)

assisted by

CANON H. E. HONE (Rector of Barnsley)
Rev. GEO. E. JOHNSON (Mayor's Chaplain)

and the

GRIMETHORPE COLLIERY BAND

" Lest We Forget "

E. Cheesman Ltd., Printers, Barnsley.

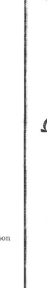

Lest We Forget: back and front pages of the programme from the United Memorial Service in respect of the fifty-eight miners killed in an explosion at Wharncliffe Woodmoor 1, 2 and 3 Colliery (where my father worked), August 1936. (*NUM*)

Young Barnsley miner Ron Palmer, in a photographer's studio in Blackpool on a day out, 1937. A year earlier, just sixteen years old, Ron was woken up at his Barnsley home and told to report to Wharncliffe Woodmoor 1, 2 and 3 pit, where he worked, as there had been an explosion. His job when he arrived was to write down the names of the dead men as bodies were brought out of the pit. (*Brian Elliott*)

Men from Waleswood Colliery display their rescue equipment, including a canary cage, outside the Rotherham and District Rescue Station, *c.* 1920. (*Norman Ellis*)

Perhaps a few years later from the previous photograph, a rescue team from a Doncaster area pit show off their Proto breathing and associated equipment. The Doncaster Station was established in 1904, only two years after the first ever rescue station was opened at Tankersley. The station functioned from a handsome building off Wheatley Hall Road, now demolished. As well as servicing Yorkshire disasters at Bentley, Hatfield and Lofthouse, men from here assisted the retrieval of bodies from the Nypro chemical plant explosion in June 1994. (*Norman Ellis*)

Charles Askew in pit rescue equipment at the Prince of Wales Colliery, Pontefract, *c.* 1910. (*NUM*)

Wharncliffe Woodmoor rescue team, looking understandably bedraggled having been involved in searching underground in terrible conditions during the 1936 disaster at the pit. (*Brian Elliott*)

The NUM Yorkshire Area Executive, photographed in the magnificent setting of the Miners' Hall, Barnsley in 1958, a century after the Yorkshire Association was formed. Front row, sitting (left to right): Frank Machin (author of *The Yorkshire Miners*), T.H. Ashman (Financial Secretary), Sam Bullough (Vice-President), J.R.A. Machen (President), F. Collindridge (Gen. Sec.), J.T.E. Collins (Compensation Agent) and E. Wainwright (NEC); middle row, standing: T. Ryan (Monk Bretton), D. Smith (Denby Grange), E. Young (Treeton), J. Finnie (Thurcroft), T. Burke (Barnburgh), H. Dixon (Wentworth Silkstone), H. Dore (Bullcroft), H. Miles (Monckton), H. Lockwood (Grange Moor), A. Smithhurst (East Ardsley); back row, standing: A. Hepworth (Staff.), D. Sheldon (Highgate), B. Goddard (Houghton Main), J.D. Gray (Ackton Hall), J. Pashley (Micklefield), W. Moorhouse (Newland), G. Welsh (Elsecar). (*J.R. Roberts/NUM*)

The Old White Bear Inn, Shambles Street, Barnsley where the South Yorkshire Miners' Association was formed in 1858. This historic pub, where John Wesley preached in the 1780s, was demolished in the 1930s. (*NUM*)

Miners' children enjoy a party at the miners' offices in Barnsley, Christmas 1937, a year or so after the Wharncliffe Woodmoor disaster. (*NUM*)

Three historic examples of purpose-built Yorkshire miners' houses:

Old Row, Elsecar, built by Earl Fitzwilliam in the late eighteenth century. (*Brian Elliott*)

Lundhill Row, near Wombwell, built to serve Lundhill Colliery and Cortonwood, mid-nineteenth century. (*Brian Elliott*)

Silkstone Row, Altofts, *c.* 1904, built in the late 1860s when Pope & Pearson developed the area for mineworkers' families. There were fifty-two three-storey houses with rear yards in this remarkable terrace. (*Old Barnsley*

A mass picket at Saltley Gate, Birmingham during the 1972 miners' strike. It was a landmark victory for the miners, the works having to close. The Tory government set up the Wilberforce Inquiry which resulted in a significant pay settlement. (*NUM*)

The saddest moment of the 1972 strike was the death of Hatfield Main picket Fred Matthews and his funeral at Hatfield, near Doncaster when an estimated crowd of 10,000 attended. Fred was crushed to death by a lorry when he was picketing outside Keadby Power Station. (*Brian Elliott*)

The 'day shift' picketing Rockingham Colliery, 28 February 1972. Edward Magon is on the extreme left and Ken Vickers on the extreme right. (*A.K. Clayton*)

The 'afternoon shift' at Rockingham Colliery during the 1972 strike, 1 March. (*A.K. Clayton*)

Redbrook (Dodworth) miners backing their Union, supporting strike action for improved wages, in 1974. In Yorkshire the poll was estimated to be 80–90 per cent in favour, with some pits almost unanimous for action, according to the *Yorkshire Post*. (*NUM*)

Beryl Robinson, an administrative worker employed at the Yorkshire NUM offices in Barnsley, displays a tie that she designed in commemoration of the 1972 'Battle of Saltley Gate', described by the Union as 'a turning point in the strike'. (*NUM*)

Groups of pickets face a large contingent of police, the latter arranged in a military formation, in a field near British Steel's Orgreave Coke Plant, on a key day during the 1984–85 strike, 18 June 1984. (*Arthur Wakefield*)

Miners outcropping for coal off Broadway, Barnsley, during the 1984–85 strike. (*Brian Elliott*)

A rare photograph (taken on 25 January 1985) showing pickets playing cards inside 'The Alamo' hut, built outside the entrance to Cortonwood Colliery, a place that attracted a great deal of media interest throughout the 1984–85 strike. The man wearing a black hat is regular picket Terence Picken who had just come out of hospital after contracting pneumonia. (*Daily Express/ Terence Picken*)

An exterior view of 'The Alamo' at Cortonwood. The picket hut got its name from the site of the 1836 siege in San Antonio, Texas during the struggle for independence against Mexico which featured a handful of volunteers, including the legendary Davy Crockett, all of whom lost their lives. (*Arthur Wakefield*)

A minor's message in protest against the Conservative government's pit closure programme is brought to the streets of London by 11-year-old Matthew Hancock, Wednesday 21 October 1992. Matthew's father, Ken, was Union Branch Secretary at Grimethorpe Colliery. (*Jim Moran*)

After the 1984–85 strike, Cortonwood Colliery was one of the first to face closure and the pit buildings were rapidly eradicated from the local landscape. It was the proposed closure of Cortonwood that was one of the main sparks for the 1984–5 strike. The photographs below are part of a small series taken at the colliery, shortly before clearance of the site, in April 1986. (*Brian Elliott*)

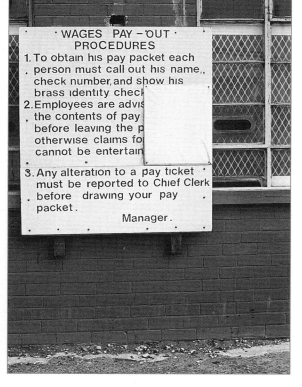

South Elmsall miner
Arthur Wakefield in
fancy dress, collecting
for the Frickley Ladies
Against Pit Closures
Group at the Notting
Hill Carnival, in London,
Sunday 26 August
1984. This photograph
appeared on the front
page of the *Guardian* the
next day. (*Martin Argles*)

An example of one of the many Miners' Strike plates produced to commemorate the longest mass dispute in trade union history. The passage on the front has the concluding phrase, 'The unity and solidarity created in the 1984 struggle shines like a beacon lighting the way for a more just society.' (*Brian Elliott*)

A limited edition plate produced in 1995 to commemorate the tenth anniversary of the 1984–85 miners' strike. It features the three key pits that triggered the start of the dispute: Cortonwood (Yorkshire), Polmaise (Scotland) and Snowdon (Wales). Plans are currently being made to commemorate the twentieth anniversary. (*Brian Elliott*)

When history repeated itself: scenes from the re-enactment of the 'Battle of Orgreave', described as 'The English Civil War Part Two', an art event created by Jeremy Deller and organised by Artangel, 18 June 2001. A film of the event was broadcast on Channel 4 television in 2002. (*Brian Elliott*)

Many Orgreave veterans took part in the re-enactment event. Here is a group assembled for the filming. (*Jeremy Deller*)

Five former 'Flying Pickets' return to the site of Silverwood Colliery, near Rotherham, in June 2003, nineteen years after the 1984–85 miners' strike. Left to right, they are: Bob Wilson, Bob Taylor, Darren Goulty, Bruce Wilson and Shaun Bisby. (*Brian Elliott*)

4

Proud Parades
& Grand Galas

The Church Lane (Redbrook) branch of the Yorkshire Area NUM display their silk banner at the 1973 miners' demonstration, held in Wakefield. The gentleman wearing the tie is Ron Palmer. (*Ron Palmer*)

PROGRAMME

OF THE BUSINESS AT THE

Demonstration

MEETING,

TO BE HELD IN THE

ENDCLIFFE PARK, SHEFFIELD,

ON

Monday, June 21st, 1909,

AT 1-30 O'CLOCK P.M.

————>◆◆◆<————

No. 3 Platform.

Barnsley hosted a series of Yorkshire miners' demonstrations during the late nineteenth and early twentieth centuries. This undated example shows a large assembly in one of the market areas, a policeman on a white horse prominent. (*NUM*)

Programme (no. 3 platform) for the 1909 Yorkshire miners' demonstration, held in Endcliffe Park, Sheffield. It had been held here in 1888 but with Sheffield seen as 'at the fringe of the Yorkshire coalfield' (*Sheffield Daily Telegraph*) it was perhaps not surprising that this occasion was the last time that the annual demonstration was held here. (*NUM*)

Keir Hardie, the cloth-capped MP for Methyr Tydfil, was the principal speaker in 1909. He was a former miner who had led the strike in the Lanarkshire coalfield in 1880 and helped to found the Independent Labour Party twelve years later.

Hardie was not afraid to use the occasion to draw attention to the abysmal housing conditions that working class people were experiencing in the locality: 'It is a very nice park that we are in, but there are parts of Sheffield that are far from being nice. I entered Sheffield by way of Parkgate, and if the approach to Hell is any more hideous I am sorry for our old friend the Devil.'

The reporter from the *Sheffield Daily Telegraph* described the busy and colourful scene but was critical of the brass bands:

Sheffield turned out in great force to see the parade from Angel Street to Endcliffe Park where the speeches took place. What they saw was a sight familiar in other towns which are more closely connected with colliery work. Sturdy miners marched along in an apparently unending stream, and by their sides marched many wives and children. The annual demonstration is as much the women's day as the men's and there were almost as many women as men in the procession.

In addition to dressing in their Sunday best, many of the miners wore sashes and favours of many colours. The banners had a massive solid look, but the most massive thing of all was the music. Far be it from us to criticise those brass bands that blared away, in all manner of march tunes. A succession of sixty brass bands is too much for the ordinary citizen, but as we all know the miner likes his music as he likes his Association – strong.

Although the Demonstration, so far as many of the miners are concerned, ends with the procession, there were crowds of people in front of the three speakers' platforms to hear the resolutions and speeches

The first platform was the most attractive of the three. People assembled round it more readily than they did the others, and this crowd was always the largest.

No miners' demonstrations were held between 1912 and 1931 but branches did display their banners in their communities on appropriate occasions. Here, for example, are officials from the Darfield Main branch in Wombwell, perhaps during the 1926 strike. Notice the sashes worn across the Sunday-best suits, several of the men even sporting bowler hats. (*R.J. Short*)

Exceptionally, the 1938 gala was cancelled due to the sudden death, in his Yorkshire NUM office, of Herbert Smith who had been President since 1906. The fourth annual demonstration after the Second World War took place in Rotherham in 1950. Here we can see the Rotherham Main (no. 2 branch) in the parade, at Canklow. (*Rotherham Archives and Local Studies Library*)

Part of the silk Aldwarke banner may be seen on the right of the photograph, probably at the assembly point (in Rotherham?). (*Rotherham Archives and Local Studies Library*)

'Coal Sunday' was celebrated in Hoyland, at the Princess Theatre, when MFGB President, Will Lawther spoke on behalf of the Yorkshire miners. Details on the poster concerning the parades suggest that a good number of banners would have been seen, with an invitation to all miners 'irrespective of the colliery in which they work' to join one of the two parades. (*Brian Elliott*)

HOYLAND COAL SUNDAY

OCTOBER 10th, 1943

MEETING

in the

PRINCESS THEATRE
HOYLAND

at 10.45 a.m. Speakers:—

Mr. Will Lawther

President of Mineworkers' Federation of Great Britain,

Lord Winster

Formerly Commander Fletcher, M.P.

PARADES.

Rotherham's last pit pony is paraded through the streets of the town in 1957 or 1958. Before the Second World War, ponies were widely used in the industry, 38,000 in Durham alone. By 1992 there were only two small mines still using ponies, one of them at Nant-y-Cafn, West Glamorgan, hauling wagons in an opencast mine. (*Rotherham Archives and Local Studies Library*)

A marvellous oblique aerial view of a procession of Yorkshire branches and their banners, probably taken from the top of the Gaumont cinema, Doncaster, turning into Hallgate, on Saturday 15 June 1957. In the foreground is the Altofts branch, followed by Pontefract's Prince of Wales, while just visible in the distance is Fryston and Dinnington. (*NUM*)

Another fine view of the 1957 procession in central Doncaster, a marching brass band preceding the Dinnington branch contingent. (*NUM*)

of Commons, as a man who is always out for the best for those he represents. He has never stinted himself in order to get it. If there is some point of difficulty in the mining industry we only have to make a request and Arthur comes down and explains very quickly the whole circumstances, and that is why we have come to like him so much, and that is why the Union appreciates the work he has done.

We appreciate the fact that our speakers have come down for this Centenary Celebration. I am certain you all agree with me that we accord to both Harold and Arthur our warmest wishes, Arthur on his retirement, and Harold in going ahead in the political future : we accord them our very warmest wishes and the sincere greetings of the whole of the miners of this County. I second the Resolution.

(*The vote of thanks to the speakers was carried with applause.*)

The Chairman : Ladies and gentlemen, I hope you will stay for a little ceremony we now have to carry out. When there are any functions of this kind in Yorkshire we like if possible to at least give some little memento of the occasion, and in this case we hope it will carry with it happy memories despite the miserable circumstances of the weather. It gives me very great pleasure, Mr. Wilson, to present to you this miniature miner's lamp as an emblem of the coalmining industry. Further, we are disappointed at the absence of Mrs. Wilson, but we have not forgotten the lady, and I hope you will pass this little gift to Mrs. Wilson with the best wishes of the Yorkshire Miners; it is suitably inscribed. Also, I have here a most stirring volume of the history of the Yorkshire Miners, a beautifully bound volume, and I hope, Mr. Wilson, that in some leisure hours you will find enjoyment and instruction in it, with an insight into the happenings of the past.

Mr. H. Wilson : This is a very great thrill for me. I am very pleased indeed to get this miner's lamp as a memento of this week-end. I did not know I was to receive this beautiful volume of the history of the Yorkshire Miners. I shall be very proud to have it. I do not get much time for reading, but one kind of reading I do like is social and economic history, especially when it is wrapped in humanity. I shall read it with very great pleasure.

The Chairman : And in your case, Arthur, I do not know whether it is like sending coals to South Wales to present a similar lamp to you, but I do so in the hope it will give you happy memories of your visit to Yorkshire. You may be going into retirement, but take it from me you still look a very active chap, and I am sure we shall see you at National Conferences. On any occasion you may wish to visit this County of the Broad Acres I hope you will let us know, for you will be very welcome. Therefore please accept this memento, Arthur. And this little gift is for Mrs. Horner,

48

A special celebration took place on 17 May 1958 to mark the centenary of the South Yorkshire Mineworkers' Association. A grand reception and banquet was organised in Barnsley, using Fretwell Downing of Sheffield as caterers, for at least 500 guests, housed in a decorated marquee. (*NUM*)

The 1958 Centenary Gala, scheduled for Locke Park, Barnsley on Saturday 21 June was a complete wash-out, but the speeches were heard in the Miners' Hall and Arcadian Hall. J.R.A. Machen chaired the meeting in the Arcadian Hall, introducing the Mayor of Barnsley (Councillor G. Skelly), Comrade Komogortsev (representing Soviet miners), Hugh Gaitskill, MP (Labour Leader) and W.E. Jones (President, NUM). The vote of thanks was given by J.T.E. 'Eddie' Collins and Dave Griffiths, the Rother Valley MP. In the Miners' Hall, Chairman Mr F. Collindridge introduced Harold Wilson, MP, A.L. Horner (Gen. Sec. NUM), Sam Bullough (Vice-President, Yorkshire Area NUM), and W.T. Paling, the Dewbury MP. This extract from the published proceedings relates to presentations made to Harold Wilson in the Arcadian Hall. (*NUM*)

THE
YORKSHIRE
MINERS

A HISTORY
by
FRANK MACHIN

Volume 1

Frank Machin's book, *The
Yorkshire Miners, A History*
starts in 1858 and continues
until the merger with the West
Yorkshire Association in 1881.
Unfortunately, the author's
premature death meant that
a second volume did not
appear until Carolyn Baylies'
commission, forty years later.
(*Brian Elliott*)

An interesting scene from the 1959 Yorkshire Miners' Gala and
Demonstration, held in Wakefield. A small child rides on the old Bentley
Colliery Branch banner which commemorates the NUM foundation in
1946 as a 'New Day' dawning, proclaiming 'Unity is Strength'. (*NUM*)

The new Bentley
banner, seen here
during the 1984–85
strike, is one of
several that feature
the Yorkshire Miners'
(and later national)
President, Arthur
Scargill. It was
completed in 1980
and cost £1,800,
supported by a small
members' levy. The
portrait is by Margaret
Burlton. The obverse
shows scenes from
the Bentley disasters
of 1931 and 1978.
(*NUM*)

More street scenes from the 1959 Wakefield demonstration. Recognise the principal lady speaker, seen in the first photograph? It was a young Barbara Castle, MP for Blackburn, and new Chairman of the Labour Party. Her published diaries remain an important source of information for the subsequent Wilson era. (*NUM*)

Opposite above: A brass band leads the way for the Elsecar Branch and their fine banner on the march through Rotherham in 1960. The banner displays the legend: UNITED TO ASSIST EACH OTHER. Following on is the contingent from Aldwarke Main whose banner extolls DIGNITY AND RESPECT. Note the size of the crowd lining the street. (*NUM*)

Opposite below: In 1963 (15 June) the Yorkshire Miners' Gala returned to Doncaster. This interesting photograph shows some of the various branches gathering together at the traditional assembly point which was Glasgow Paddocks (behind the NCB offices, Catherine Street), prior to the parade through the town towards the racecourse. Recognisable branches include Glasshoughton, Wheldale, Prince of Wales and Fryston. (*NUM*)

This wonderful photograph from the 1963 Doncaster Gala shows the Askern Branch banner, led by a veteran miner. It was traditional for young children, in this case five girls, to 'ride the banner'. (*NUM*)

Members of the Thurcroft Branch, with FROM OBSCURITY TO RESPECT on their banner, stride out towards Doncaster's famous racecourse, 15 June 1963. (*NUM*)

Setting the pace: the leading marchers begin the walk to Doncaster Racecourse in 1963. At the front (left to right) is J.T. Leigh, F. Hayday, Sam Bullough, D.H. Davies, Fred Collindridge and Sydney Schofield. Just in shot, at the back, on the left, is Barnsley MP Roy Mason, soon to become a junior minister. The main speaker was the Labour Leader, Hugh Gaitskill. (*NUM*)

Two of the Wharncliffe Woodmoor (Carlton, near Barnsley) branches march along South Parade, Doncaster, during the 1963 Yorkshire miners' gala. Joe Hall, featured on one of the banners, was a most popular miners' leader and hero. A Yorkshireman, working at Darfield Main at the age of twelve, he was a brilliant orator, known with affection quite simply as 'Our Joe'. He gained tremendous respect in Yorkshire and elsewhere for his supportive action to save lives during mining disasters such as Barnburgh, Bullcroft, Barnsley Main, North Gawber, Wharncliffe Woodmoor and Gresford (North Wales). (*NUM*)

Sam Bullough, addressing a large crowd at the racecourse during the 1963 Doncaster Gala. For many years a respected figure, he was Vice-President of the Yorkshire Miners. (*NUM*)

The Yorkshire Miners' Coal Queen contest was always a popular part of the Gala. Miss Carol Lycett won the award at the 1963 event. (*NUM*)

The Fancy Dress competition was also a traditional event at modern miners' galas. Here we can see 'Prince Monolula' (dressed as himself! and holding the sporting pink) apparently winning first prize at the racecourse in 1963. Many people will remember 'Prince Monolula' at Doncaster Races during the 1950s and early 1960s when 'being a black man' at Yorkshire sporting events was something of a rarity. (*NUM*).

Opposite above: The 1964 Yorkshire Miners' Gala and Demonstration was held in Castleford, the various branches (e.g. Maltby Main, Firbeck, Bower Unit, Thurcroft) seen here are in the assembly area. (*NUM*)

Opposite below: Another marvellous photograph showing children riding on on the banner, Castleford, 20 June 1964 (*NUM*)

Above: An attractive distant view of the crowd, speakers' platform and boxing ring, in the park at Castleford, June 1964. (*NUM*)

Right: Anthony Greenwood (left) MP and Chairman of the Labour Party was one of the main speakers at the 1964 Castleford gala. Here, he gratefully receives from Sam Bullough the traditional present of a miners' lamp. (*NUM*)

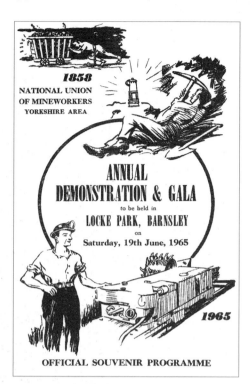

Front cover of the 1965 gala programme. (*NUM*)

Opposite above: The gala procession was always a happy family occasion, as may be seen by the smiling faces from the Manvers Main branch, photographed at the start of the 1965 Gala. (*NUM*)

Opposite below: The Armthorpe branch banner and brass band 'compete' with a religious placard during the procession to Locke Park, Barnsley in 1965. (*NUM*)

The leading group starts the procession to the assembly in Locke Park, Barnsley, at the start of the 1965 Yorkshire Miners' Gala. They were preceded by the West Riding Fire Service Silver Band. At the front, left to right: Sidney Schofield, Lord Collinson (Chairman of TUC, guest speaker), Mayor and Mayoress (Alderman and Mrs A. Butler), Sam Bullough, George Brown MP (First Secretary of State, guest speaker), Mrs Brown, Roy Mason, MP, and J.T. Leigh. (*NUM*)

Competitors for the Yorkshire Coal Queen contest face the judges at the 1966 Gala in Clifton Park, Rotherham. (*NUM*)

One of the main speakers in 1966 was Richard Marsh MP, Minister of Fuel and Power, seen here (at the centre of the photograph) enjoying a cup of tea courtesy of two picnicking miners. Marsh was dismissed by Harold Wilson in 1969. He resigned his seat in 1971, becoming chairman of British Railways, and was eventually a supporter of Margaret Thatcher. (*NUM*)

A delightful photograph showing Prime Minister Harold Wilson greeting children assembled in Thornes Park, Wakefield, during the Yorkshire Miners' Gala, 17 June 1967. (*NUM*)

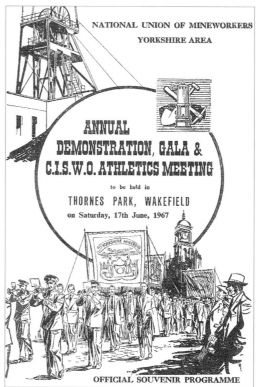

Harold Wilson, after being presented with a miners' lamp, making a point during his address to the 1967 Gala in Thornes Park, Wakefield. A year earlier he had led the Labour government to a convincing election victory of almost a hundred seats. (*NUM*)

The front cover of the 1967 Yorkshire Miners' Gala was specially designed for the occasion, the event also including an athletics meeting. In future years the programmes were plain, with minimal graphics and therefore far less interesting. (*NUM*)

This superb aerial view shows Yorkshire miners' branches assembling at Glasgow Paddocks, Doncaster before the start of the 1968 Gala. A woman walks on the road, a placard attached to her proclaiming 'Equal Pay for Women'. (NUM)

The guest speaker in 1968 was George Woodcock (centre of the photograph), in his penultimate year as General Secretary of the TUC. Generally regarded as a moderate, he was a quietly spoken, very able and popular union leader. Woodcock's final period of office coincided with Barbara Castle's 'In Place of Strife' proposals, opposed so much by the Labour movement, including key members of Wilson's cabinet. (*NUM*)

The speaker representing the miners at the 1968 Doncaster Gala was the highly regarded General Secretary of the NUM, Will Paynter, seen here at the rostrum. (*NUM*)

A marvellous photograph showing the march down Market Hill, Barnsley during the 1969 Yorkshire miners' gala. At the forefront is the Grimethorpe Branch banner, featuring the great J.A. 'Joe' Hall, closely followed by the Houghton Main contingent. Note the packed crowd by the market stalls and on the pavement at the Eldon Street junction. Market Hall had a lot of character in the 1960s. (*NUM*)

'Just Like Mi Dad': a lovely photograph of the children's fancy dress competition, held in Locke Park as part of the 1969 Gala festivities. (*NUM*)

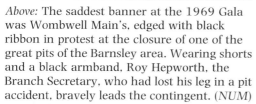

Above: The saddest banner at the 1969 Gala was Wombwell Main's, edged with black ribbon in protest at the closure of one of the great pits of the Barnsley area. Wearing shorts and a black armband, Roy Hepworth, the Branch Secretary, who had lost his leg in a pit accident, bravely leads the contingent. (*NUM*)

Above right: Roy Mason, MP, makes an animated speech in Locke Park at the 1969 miners gala in his home town of Barnsley, when he was Minister for Fuel and Power. A former miner at Wharncliffe Woodmoor 4 and 5 pit, Roy was elected as the town's MP at the young age of twenty-nine, in 1953, and continues to be an active member of the 'Second House' as the Lord Mason of Barnsley. (*NUM*)

Right: Lawrence Daly, Secretary of the NUM, speaking in Locke Park at the 1969 Yorkshire Miners' Gala. Daly came to prominence as a leader of the Scottish miners, and often a powerful and popular speaker. (*NUM*)

In 1970 the annual Yorkshire Miners' Gala moved to Pontefract. Here, on 13 June, the procession of branches and banners march through the town, crowded with spectators. (*NUM*)

Principal speaker in 1970 was the TUC leader 'Vic' Feather who, as will be obvious from his full name (Victor Grayson Keir Hardie Feather), was a born socialist. He served as General Secretary during a very troubled period and had the reputation of being an excellent negotiator. (*NUM*)

Two children dressed as 'Health and Safety' catch the eye of the photographer, during the 1970 Pontefract gala. (*NUM*)

Arthur Scargill with NUM President Joe Gormley (centre) and General Secretary of the Labour Party, Bill Simpson (right) at the 1972 Yorkshire Miners' Gala in Thornes Park, Wakefield. (*NUM*)

Arthur Scargill speaking at the Wakefield gala in June 1973. A former Woolley Colliery miner, the new President of the Yorkshire Area came to national prominence during the 1972 strike, when, in his mid-thirties, he successfully deployed mass pickets, notably at Saltley Gate, Birmingham. The Gala was held only three months after the tragic Lofthouse disaster when seven miners were killed. Arthur Scargill represented the NUM at the subsequent inquiry. (*NUM*)

The Edlington (Yorkshire Main) Branch walk towards Doncaster Racecourse during the Yorkshire Miners' Gala, 15 June 1974. (*NUM*)

Tony Benn (left) receives a painting from Yorkshire artist Ashley Jackson, with Arthur Scargill assisting. Ben was the Secretary of State for Energy, and main speaker at the 1976 Yorkshire Miners' Gala, held in Wakefield. (*NUM*)

Barnsley was the venue of the the 1983 and 1987 Yorkshire Miners' galas. Here we can see the shirt-sleeved Dennis Skinner, MP for Bolsover between Peter Heathfield (left) and Arthur Scargill (right), making their way to Locke Park. (*Philip Thompson/NUM*)

The leading party walk through Rotherham during the 1985 Yorkshire Miners' gala. On the left (with his dog) is the singer and entertainer Mike Harding, behind him Tony Benn, then Arthur Scargill (light suit), Owen Briscoe, Jack Taylor (Yorkshire Miners' President), Sam Thompson (Yorkshire Miners' Secretary) and (extreme right) Ken Homer (Vice-President of the Yorkshire Miners). (*Philip Thompson/NUM*)

A good view of the Hatfield Main Branch banner, featuring Keir Hardie and A.J. Cook, Yorkshire Miners' Gala, Rotherham, 1985. A small boy holds a placard calling for the reinstatement of sacked miners. Hatfield is one of the few Yorkshire pits still open, under private ownership. (*Philip Thompson/NUM*)

Right: The North Gawber Branch banner at the Rotherham gala in 1985. (*Philip Thompson/NUM*)

Opposite above: Placards carried in memory of John Little and Joe Green precede a Durham banner at Clifton Park, Rotherham, 15 June 1985. (*Philip Thompson/NUM*)

Opposite below: The Cortonwood Branch banner at the Rotherham Yorkshire Miners' Gala in 1985. (*Philip Thompson/NUM*)

Right: The Royston Drift banner, featuring Arthur Scargill, in Rotherham, 15 June 1985. Royston Drift was a new mine, opened in 1976, developed as part of the NCB's Plan for Coal which anticipated increased demand during the 1980s. Sited across from the old Monckton colliery complex, it broke the world record for coalface production but by the late 1980s its reserves were exhausted and the short-lived but highly efficient pit closed. (*Philip Thompson/NUM*)

Above: Woolley miners return to work after the Great Strike, March 1985, a scene replicated throughout most mining areas. (*Brian Elliott*)

Right: Rodney Bickerstaffe, the former NUPE General Secretary, who did a great deal to support the miners during the 1984–5 strike, is presented with a commemorative miner's lamp by Arthur Scargill, in the latter's office, March 2002. Bickerstaffe had delivered an excellent lecture in the Miners' Hall and the author was invited to photograph the presentation. A few months later, in August, Arthur Scargill retired as President of the NUM, but remains as Honorary President, union matters now in the hands of two elected officials: Ian Lavery (Chairman) and Steve Kemp (General Secretary). (*Brian Elliott*)

Opposite above: The new Goldthorpe Branch banner, paraded through the streets of Rotherham during the 1985 gala. Made during the 1984–5 strike, the face features strike committee branch members Barry Miller, 'Daz' Farr, Joe Gaskill, Phil James and Frank Calvert, with clenched fists raised in salute and carrying a placard proclaiming 'NO SURRENDER'. (*Philip Thompson/NUM*)

Opposite below: The obverse of the new Goldthorpe Branch banner, used during a CND demonstration in July 1985. It shows miners using shovels to drive away rats whose faces represented Prime Minister Margaret Thatcher, NCB Chairman Ian Macgregor and Energy Secretary Peter Walker. On the photograph (left to right) are Barry Kelly (branch President), Phil James (delegate), Monsignor Bruce Kent (General Secretary CND) and Pat Miskill (committee member). (*Brian Elliott*)

A tired ex-Frickley miner takes a rest in Doncaster, at the end of the what was probably the last great Yorkshire Miners' Gala Demonstration, in June 1994. His placard laments the closure of the 100-year-old pit and, on the obverse side, the role of Michael Heseltine, the Conservative Deputy Prime Minister, who announced the massive 1992 pit closure programme, decimating the remaining Yorkshire pits. (*Norman Ellis*)

5

Work & Play

Getting the coal in – a rare sight these days. Wearing her pinny, with sweeping brush strategically placed, Mrs Muriel uses a shovel and barrow for the task, outside Woodburn's shop, Stead Lane, Hoyland Common, in 1967. (*A.K. Clayton/Brian Elliott*)

Above: Hemingfield Victoria Football Club, probably outside their local pub, *c.* 1903. Most of the players probably worked at the small local pit (see below) or at nearby collieries such as Wombwell or Mitchell's Main (*Brian Elliott*)

Right: Hemingfield Colliery's unusual headgear meant that it was known as the 'bicycle pit'. This small, almost forgotten colliery was sunk in 1840 and ceased production in 1920. The distinctive headgear was removed in 1935. (*Frank Williamson*)

Opposite: Yorkshire miners demonstrate the traditional game of knurr and spell, *c.* 1890. The sport was popular in the north of England from the mid-nineteenth century. The spring-loaded device on the ground (known as a trap or spell) ejected a pot ball (or knurr) which was struck by the bulbous end of a carved wooden club, about the length of a pick handle. The distance was then measured. A great deal of skill as well as strength was required. After the First World War, players such as Joe Edon of Barnsley and Jim Crawshaw of Stocksbridge were famous players. In the 1930s a crowd of 4,000 watched Jim beat Henry Mollart of Grenoside at Wharncliffe Side, for the world championship. At the Queens Ground, Barnsley (now the site of the Metrodome), players could compete for a prize of £100, many times a collier's wage, and betting took place. (*A.K. Clayton/Brian Elliott*)

New Stubbin Colliery, situated near to Rawmarsh, was sunk under the authority of Earl Fitzwilliam, somewhat unusually, during the First World War, as a replacement for his almost exhausted Top Stubbin pit. The 6ft-thick Parkgate seam was reached at a depth of 287yd via two shafts whose headgears are visible on this photograph, probably dating from the early 1920s when coal production was in full swing. The Thorncliffe seam was developed here in the 1930s and, after nationalization and modernisation, the famous 'Silkstone' was exploited, until it became so thin as to be too uneconomic for production, closure coming in 1978. (*Old Barnsley*)

New Stubbin Colliery FC was formed in 1921, apparently at the suggestion of Earl Fitzwilliam, whose son, Viscount Milton, provided a sports ground and pavilion. They were champions of the Sheffield Amateur League, in the 1935–36 season. Back row (left to right, players only): J. Cooper, R. Harris, G. Hillman, E. Hobson, J. Wiley, T. Cooper. Front row: W. Hammond, G. Smith, W. Hibbert (Capt.), G. Roddison, A. Cosney, A. Crossley (*Brian Elliott*)

Wharncliffe Woodmoor Colliery (Carlton, near Barnsley) cricket team in 1925. Back row, left to right, they are (players only): G. Owen, T. Fletcher, A. Blakey, J. Helliwell, W. Chambers, R. Goodall. Middle row, seated: S. Fox, A. Jones, W. Clarkson. Front row: Clarence Cooke, Walter Cooke, L. Crawshaw, F. Briggs. (*Alan Cooke*)

A view of Wharncliffe Woodmoor 1, 2 and 3, taken by the author from the colliery muckstack shortly after the pit closed in August 1966. Some of the outbuildings were subsequently adapted for commercial premises. (*Brian Elliott*)

Keeping pigeons on an allotment was a popular hobby in mining communities, and a serious sport for my grandfather, seen here, 'sat on his haunches' at his Carlton pigeon cote, *c.* 1946. Fred Elliott's homing birds took part in many long-distance races. He worked at Dodworth and Monckton Main collieries. (*Brian Elliott*)

Members of the new Royston and District Canine Society, in 1928, two years after the General Strike. An article in the local press by show secretary Leonard Ashton (sixth from right in photograph) lamented that it was very difficult for miners to exhibit their dogs because of the preparation time and expense needed. The men could not afford to take time off work to exercise and train their dogs into peak condition. (*Rosalie Bailey*)

Denaby Main (St John Ambulance) Band. Colliery companies were usually keen to encourage First Aid competitions and support brass bands. Many of these men may have been affected by the disputes of 1869, 1885 and 1902/3 when the Denaby and Cadeby Colliery Company evicted workmen and their families. (*Brian Elliott*)

Silverwood Colliery Band, *c.* 1946. The band functioned from 1908 to 1968. It was also known as the 'widows' band' because colliery fatalities frequently robbed the band of its finest players. A former bandmaster, Bill Dodd of Kilnhurst, was known as 'Injun (engine) Billy' because of his powerful lungs, a blast that could be deafening! His other claim to fame was a propensity to conduct better when drunk than sober. Silverwood was one of the few bands that had music for clog-dancing. (*Brian Elliott*)

Members of the Turner families on an allotment, probably at Goldthorpe, *c.* 1912. The man in the cap (centre) is George Lavender Turner (*b.* 1885). To his left is Martha Turner, née Gray, and the man wearing a moustache (on the right) is Henry Turner. The men were local miners. (*Millie Crowther*)

Here we can see Jim Richardson, from Grimethorpe, demonstrating another very popular old collier's game – nipsey – to a group of children from Milefield Primary School in the late 1990s. Part of a pick shaft (or purpose-made stick) was used as a bat to strike a 2.5oz piece of wood which was placed on a house-brick and flipped into the air. The local distance champion was George Roystone who achieved a phenomenal 200yd in the 1920s. (*Brian Elliott*)

A cyclist takes a composed rest at Kiveton Park Colliery in about 1910, which, as can be seen, had excellent railway connections. Coal from the Barnsley seam was extracted from the mid-1860s. The Thorncliffe, Silkstone, High Hazel and several other seams were worked from the 1890s. Production ceased in 1994 but no buyer could be found for the pit. I remember walking round the fine old colliery offices, with their distinctive clock tower, during the late 1990s when the listed building was being transformed into an educational and training centre. The substantial pithead baths, built to the design of W.A. Woodland in 1938, are even worthy of mention in Pevsner's *Yorkshire West Riding* edition of the *Buildings of England* books. (*Old Barnsley*)

Railways are also prominent in this fine photograph, *c.* 1920s, of Dinnington Colliery by Edward Leonard Scrivens. When the pit was developed in the early 1900s the sinkers were housed in make-do accommodation in an area known as 'Tin Town'. Soon, model housing was built, and in 1928 a modern Mining and Technical College was opened by Viscount Chelmsford. By 1992 the pit had closed and soon there was little left on the local landscape to remind us of its former presence. I recall ex-Dinnington miner Eric Botherton (*b.* 1919) calling to see me at my Creative Writing class at Rother Valley College, Dinnington. A few years earlier, in 1994, he had completed a brief history of the village which included an account of his life at the pit, including his fascinating experiences as a member of the rescue team. (*Old Barnsley*)

Mr Bullock, just off his shift at Hickleton Main, meets his sons Alan and David and their pet ex-pit pony 'Bob' on Green Lane. Note the 'like my dad' clogs worn by Alan. (*Peter Davis*)

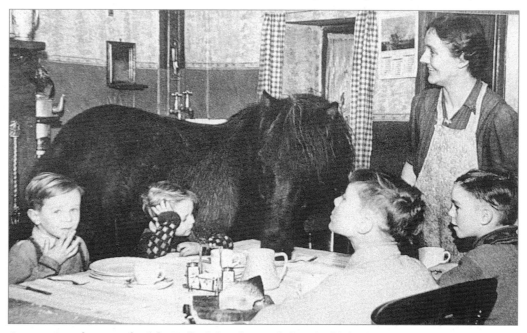

An amazing photograph of the Bullock family – including 'Bob' the former pit pony! – having tea at their home, Furlong House, Green Lane, Barnburgh. (*Peter Davis*)

6

Monuments & Memorials

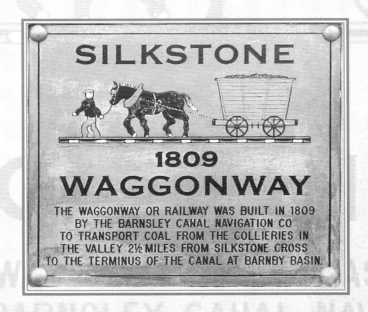

Plaque in commemoration of the old Silkstone wagonway of 1809, at Silkstone crossroads. Silkstone coal was regarded as the most lucrative of the Yorkshire seams during the nineteenth century.

The building of a feeder railway was a vital link between the developing pits in the Silkstone valley and the Barnsley Canal extension at Barnby Basin. It was authorised by Act of Parliament on 28 March 1808 and work commenced in January 1809. An account of the 'Silkstone Railway' by John Goodchild can be seen in *Aspects of Barnsley 2*, edited by Brian Elliott (Wharncliffe Books, 1994). (*Brian Elliott*)

 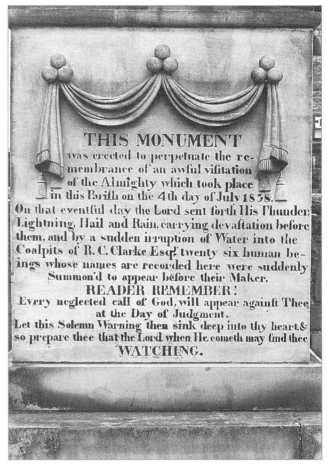

Above left: Of all the mining monuments extant in Yorkshire, the saddest must surely be the large obelisk in Silkstone churchyard, marking the remains of the twenty-six children (including eleven females), aged from 7 to 17, who drowned in the Husker pit disaster of 1838. The story of the disaster and its widespread aftermath have recently been explored in Alan Gallop's excellent book, *Children of the Dark: Life and Death Underground in Victoria's England* (Sutton Publishing Ltd, 2003). (*Brian Elliott*)

Above right: Detail of the inscription to the Husker pit monument, placed there under the instructions of the colliery owner, R.C. Clarke. Note the Biblical overtones and 'Solemn Warning' to other mortals. (*Brian Elliott*)

Opposite above: A superb reconstruction of a 'Silkstone wagon' can be seen at Silkstone Crossroads. The main coal owners taking advantage of the new railway were the Barnsley attorney Jonas Clarke of Noblethorpe Hall, Samuel Thorp of Banks Hall, the Wilsons of Barnby Basin and Popplewell & Wilkinson. (*Brian Elliott*)

Opposite below: Many of the stone sleeper blocks of the wagonway are still *in situ*, marking the course of the historic routeway. The development of steam railways meant that traffic gradually declined from the 1840s, and the rails were removed in about 1871–2. (*Brian Elliott*)

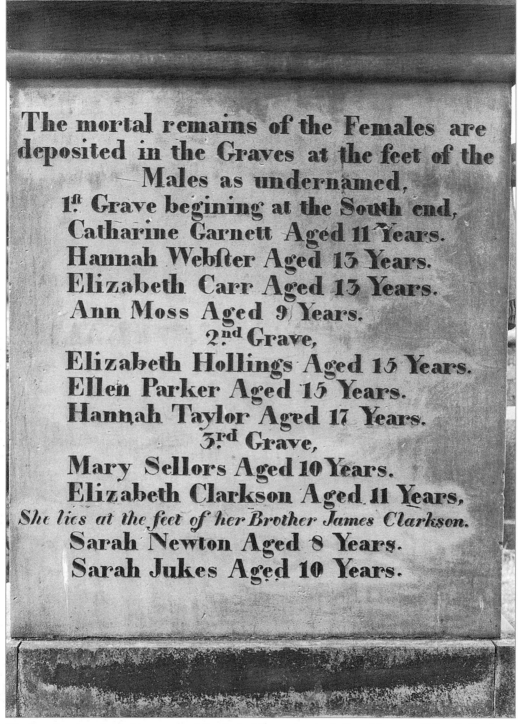

The mortal remains of the Females are deposited in the Graves at the feet of the Males as undernamed,

1st Grave begining at the South end,

Catharine Garnett Aged 11 Years.
Hannah Webster Aged 13 Years.
Elizabeth Carr Aged 13 Years.
Ann Moss Aged 9 Years.

2nd Grave,

Elizabeth Hollings Aged 15 Years.
Ellen Parker Aged 15 Years.
Hannah Taylor Aged 17 Years.

3rd Grave,

Mary Sellors Aged 10 Years.
Elizabeth Clarkson Aged 11 Years,
She lies at the feet of her Brother James Clarkson.
Sarah Newton Aged 8 Years.
Sarah Jukes Aged 10 Years.

The females were buried in graves at the feet of the males, including 11-year-old Sarah Clarkson who was placed at the feet of her 16-year-old brother James. (*Brian Elliott*)

The frequency and extent of mining disasters increased from the 1840s, with the coming of the railways and the development of large-scale mining in the Barnsley area. At the Lundhill Colliery on a cold February day in 1859, an explosion of firedamp resulted in the deaths of 189 men and boys. The Kellett family lost five sons. An estimated crowd of 15,000 assembled at the pit-head, including 'coachloads of excursionists', and artists were dispatched from London by the *Illustrated London News*. The now black obelisk in a far corner of Darfield churchyard is a grim reminder of this terrible event. (*Brian Elliott*)

Detail of the Lundhill disaster monument. As at Husker nineteen years earlier, a Biblical warning completed the inscription. Just imagine the scenes in Darfield churchyard when so many interments were taking place during the last few days of February. Two years later a party of literary figures, including Charles Darwin, toured the infamous pit. (*Brian Elliott*)

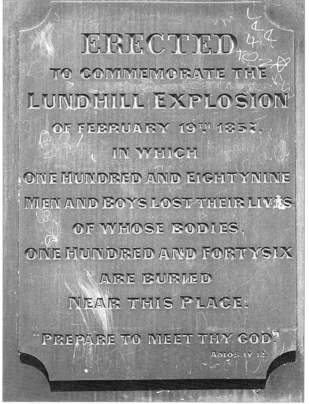

ERECTED BY PUBLIC SUBSCRIPTION
TO THE MEMORY OF THE 143 MEN AND BOYS
WHO WERE KILLED BY AN EXPLOSION
IN SWAITHE MAIN COLLIERY
ON DECEMBER 6TH 1875.
"IN THE MIDST OF LIFE WE ARE IN DEATH."

Opposite: Another blackened and almost forgotten monument to a Victorian coal-mining tragedy can be seen in front of St Thomas's church, Worsbrough Dale, lamenting the occasion when Swaithe Main pit 'fired' and 143 men and boys perished. Once again, a stern religious phrase warns of impending death but was of little consolation to so many grieving families in 1875. (*Brian Elliott*)

Opposite inset: On the obverse of the Swaithe Main monument is a small carving of a coal tub, lamp and the miner's basic tools of pick, shovel and riddle. (*Brian Elliott*)

Right: Britain's worst mining disaster (until Senghenydd, 1913) took place in December 1866 at the notoriously fiery Oaks Colliery near Barnsley when 361 men and boys were killed in two (from a series of many) explosions. Among the casualties were twenty-six rescue workers, led by mining engineer Parkin Jeffcock. This obelisk and monument was erected near the site of the pit, on Kendray Hill, forty-seven years after the event, by a Barnsley businessman as a tribute to Jeffcock and other heroes who lost their lives. It also cites the bravery of two other volunteers, John Mammatt and Thomas Embleton, who were lowered into the stricken pit to rescue the sole survivor. (*Brian Elliott*)

Right: Many of the Oaks victims were buried in Ardsley and Monk Bretton churchyards and in Barnsley cemetery, though about a hundred bodies were never recovered. This almost forgotten monument in Ardsley churchyard was erected by public conscription to commemorate the men and boys killed in the disaster and marks the burial place of thirty-five local miners. (*Brian Elliott*)

The statue known as *Gloria Victus* at the top of the 'Oaks monument' underwent a detailed restoration in 1998 at Sheffield Hallam University after vandals had knocked it off its plinth and broken part of the angel's wing. Thankfully, the statue was restored to its former glory, though at considerable expense. (*Wes Hobson*)

The bronze statue of a grieving widow forms part of a fitting tribute to the men and boys who lost their lives at Cadeby and Denaby Main collieries, and also to the women 'who shared their lives and suffered their loss'. The sculpture stands in a public garden next to Conisbrough Library, Graham Ibberson completing the commission for Doncaster MBC in 1987. (*Brian Elliott*)

Behind the grieving woman is a half-buried figure of a collier, one of his hands desperately reaching through fallen debris. (*Brian Elliott*)

The commemorative pulley wheel at Denaby Main, photographed in 1989 when land reclamation was taking place through the Environmental Department of the South Yorkshire County Council. In a hundred years (to 1968) there were 203 recorded fatalities at the pit. One plaque on the wheel base pays tribute to the men and boys who lost their lives, and another recognises the work of the ex-Denaby Main pitman Jim McFarlane, who became Leader of Doncaster MBC and lectured in Industrial Studies at the University of Sheffield. (*Brian Elliott*)

The commemorative pulley wheel in St Thomas's churchyard, Kilnhurst, near Rotherham is dedicated to the men and boys who lost their lives at the local pit between 1858 and 1989. Coal-winding stopped at Kilnhurst Colliery in the mid-1950s due to an underground link with Manvers, the pit later forming part of the Manvers Complex in the mid-1980s. Production came to an end in 1988. (*Brian Elliott*)

Detail from the Yorkshire Main pulley
wheel, Edlington Lane, near Doncaster.
This pit closed on Friday 11 October 1985
and the great headgears were soon blasted
from the local landscape. Originally
known as Edlington Main, it was the
Stavely Coal and Iron Company who was
responsible for sinking and developing the
colliery, and creating model housing in
New Edlington Village. (*Brian Elliott*)

A sunken pulley wheel serves as a
memorial to the disaster at Wharncliffe
Woodmoor 1, 2 and 3 Colliery in
Athersley Memorial Park, near Barnsley.
A plaque refers to the explosion that killed
fifty-seven men instantly and another, five
days later. The unveiling party included
Barnsley's MP, Roy Mason, and Yorkshire
NUM President Arthur Scargill. (*Brian
Elliott*)

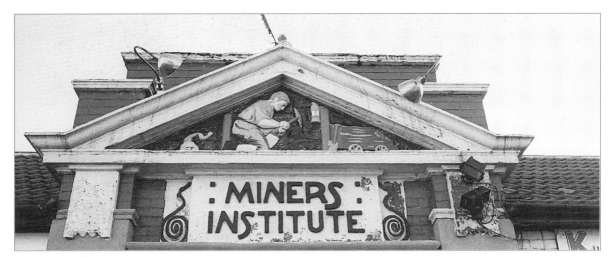

Detail from the carved pediment on the former Miners' Institute, Moorthorpe, South Elmsall. Miners reported here for duty during the 1984–85 strike. (*Brian Elliott*)

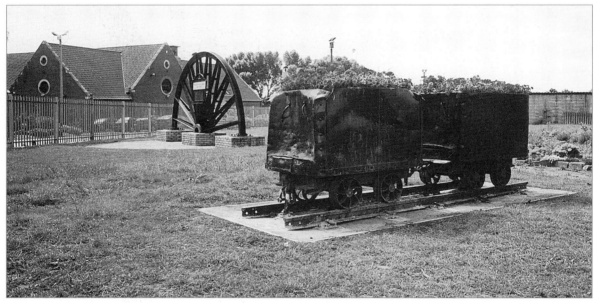

A pair of coal tubs, planted with flowers, and in the background a commemorative half pulley wheel, in the developing memorial garden at Armthorpe, near Doncaster, by the site of Markham Main Colliery. Sir Arthur Markham, who obtained the mineral rights from Earl Fitzwilliam, died shortly after sinking commenced in 1916. The Great War delayed completion, and the Barnsley Seam was not reached until 1924. Double-deck cages were used to carry tubs from the two shafts and coal-cutting machines introduced in the late 1930s, and by the 1960s over 20 per cent of output was power-loaded. The pithead baths of 1938 could cater for over 2,000 workmen. The No 1 shaft was deepened so as to access the Parkgate Seam in the early 1960s, the Dunsil Seam having been exploited a few years earlier. Accessing thinner seams would have extended the colliery's life to 2050, according to the NCB. Markham Main was one of the thirty-one selected for closure by British Coal and the Conservative Government in October 1992. Despite the intervention of a company called Coal Investments, led by ex-British Coal commercial director Malcolm Edwards, Markham closed for the final time in 1996. (*Brian Elliott*)

Detail from the Markham Main pulley wheel, dedicated to and in honour of the eighty-seven miners who were killed at the pit between 1920 and 1996. (*Brian Elliott*)

An annual remembrance service continues to take place in Arksey Cemetery in honour of the 1931 and 1978 disasters when forty-five and seven miners respectively were killed. (*Brian Elliott*)

This postcard was produced to show the scene at the mass grave for the Bentley Colliery disaster victims in Arksey Cemetery, 25 November 1931. (*Brian Elliott*)

HICKLETON MAIN COLLIERY
1892 - 1988

This plaque was erected to mark the site of
the former Hickleton Main Colliery which
was opened in 1892 and closed in 1988.

Dedicated by Barnsley Metropolitan
Borough Council and British Coal, it will
remain a memorial to all miners and their
families who suffered hardship and grief
in their pursuit of coal from this mine.

The commemoration ceremony was carried
out by the Mayor of Barnsley,
Councillor Kenneth Young
on the 6th April 1993.

Above: A small pulley wheel near the site of Hickleton
Colliery serves as a modest memorial for one of the
biggest Yorkshire pits, the new street name devised
within the context of the regeneration of the Dearne
Valley. (*Brian Elliott*)

Inset: Detail, showing the commemorative plaque on the
Hickleton memorial pulley, unveiled in a ceremony on 6
April 1993. (*Brian Elliott*)

Right: Bolton-upon-Dearne did not have its own
village pit but many men worked nearby, particularly
at Hickleton and Barnburgh collieries. Walk into the
cemetery on Furlong Road and you will see numerous
gravestones, usually provided by working colleagues,
in memory of men and boys killed in the two pits. This
particularly sad one relates to a 13-year-old lad, John
Flanagan, who lost his life in 1903. (*Brian Elliott*)

The actor Pete Postlethwaite, star of *Brassed Off* which was partly filmed in Grimethorpe, returned to the village in December 2000 to open the new Millennium Green which includes an obelisk celebrating the village's heritage. The lone cornet player is from the famous Grimethorpe Brass Band. (*Brian Elliott*)

Local schoolchildren contributed to the designs on the Millennium Green obelisk at Grimethorpe, this example featuring the village's mining history. (*Brian Elliott*)

A new pit memorial at Grimethorpe was unveiled by schoolchildren outside St Luke's church in March 2003. It carries the names of 154 dead miners killed at the pit. Father Peter Needham gave an emotional speech on the day, paying tribute to the deceased miners and their families. Excellent work in terms of fund raising and research was carried out by local people, supported by Yorkshire Forward and Waste Recycling Environmental. (*Barnsley Chronicle*)

The new granite memorial outside St Luke's church, Grimethorpe. The name of George Eastwood, my uncle, was one of the last names to be inscribed. He was killed in an underground accident that should never have happened. (*Brian Elliott*)

NUM President Arthur Scargill lays a wreath at the new monument outside the NUM offices, in March 2002, a solemn tribute that takes place every year. (*Brian Elliott*)